Quantitative Zooarchaeology
Topics in the Analysis of Archaeological Faunas

STUDIES IN ARCHAEOLOGICAL SCIENCE

Donald K. Grayson, Consulting Editor
Burke Memorial Museum
University of Washington
Seattle, Washington

Chaplin: The Study of Animal Bones from Archaeological Sites
Reed: Ancient Skins, Parchments and Leathers
Tite: Methods of Physical Examination in Archaeology
Evans: Land Snails in Archaeology
Limbrey: Soil Science and Archaeology
Casteel: Fish Remains in Archaeology
Harris: Principles of Archaeological Stratigraphy
Baker/Brothwell: Animal Diseases in Archaeology
Shepherd: Prehistoric Mining and Allied Industries
Dickson: Australian Stone Hatchets
Frank: Glass Archaeology
Grayson: Quantitative Zooarchaeology

in preparation:

Dimbleby: The Palynology of Archaeological Sites

Quantitative Zooarchaeology
Topics in the Analysis of Archaeological Faunas

DONALD K. GRAYSON

Department of Anthropology
 and
Burke Memorial Museum
University of Washington
Seattle, Washington

1984

ACADEMIC PRESS, INC.
(Harcourt Brace Jovanovich, Publishers)
Orlando San Diego New York London
Toronto Montreal Sydney Tokyo

COPYRIGHT © 1984, BY ACADEMIC PRESS, INC.
ALL RIGHTS RESERVED.
NO PART OF THIS PUBLICATION MAY BE REPRODUCED OR
TRANSMITTED IN ANY FORM OR BY ANY MEANS, ELECTRONIC
OR MECHANICAL, INCLUDING PHOTOCOPY, RECORDING, OR ANY
INFORMATION STORAGE AND RETRIEVAL SYSTEM, WITHOUT
PERMISSION IN WRITING FROM THE PUBLISHER.

ACADEMIC PRESS, INC.
Orlando, Florida 32887

United Kingdom Edition published by
ACADEMIC PRESS, INC. (LONDON) LTD.
24/28 Oval Road, London NW1 7DX

Library of Congress Cataloging in Publication Data

Grayson, Donald K.
  Quantitative zooarcheology.

  (Studies in archaeological science)
  Includes index.
  1. Animal remains (Archaeology)--Statistical methods.
I. Title. II. Series.
CC79.5.A5G7 1984   930.1'028'5   84-9201
ISBN 0-12-297280-5

PRINTED IN THE UNITED STATES OF AMERICA

84 85 86 87   9 8 7 6 5 4 3 2 1

To
J. ARNOLD SHOTWELL
friend and teacher

Contents

LIST OF FIGURES — ix
LIST OF TABLES — xiii
PREFACE — xix

Chapter 1: Introduction — 1
    Background to Two Faunal Samples — 3

Chapter 2: The Basic Counting Units — 16
    The Number of Identified Specimens — 17
    The Minimum Number of Individuals — 27
    The Relationship of NISP to MNI — 49
    The Relationship of MNI/NISP to NISP — 68
    Some Other Approaches to Counting — 85
    Minimum Numbers and Specimen Counts:
        Some Conclusions — 90

Chapter 3: Levels of Measurement — 93
    MNI and NISP as Ratio Scale Measures — 94
    MNI and NISP as Ordinal Scale Measures — 96
    Conclusions — 110

Chapter 4: Sample Size and Relative Abundance — 116

Correlations between Sample Size and Relative Abundance:
  Some Examples — 117
Exploring the Causes — 121
Conclusions — 129

Chapter 5: Taxonomic Richness, Diversity, and Assemblage Size — 131

Taxonomic Richness — 132
Rarefaction and the Comparison of Species-Abundance
  Distributions — 151
Diversity Indices and Sample Size — 158

Chapter 6: Collection Techniques, Meat Weights, and Seasonality — 168

Collection Techniques — 168
Meat Weights — 172
Seasonality — 174

Chapter 7: Conclusions — 178

REFERENCES — 181
INDEX — 199

List of Figures

2.1	Changing frequencies of *Gazella* and *Dama* at Tabūn and Wad, as calculated by Dorothea Bate from numbers of identified specimens.	19
2.2	Distributions of most abundant elements and the effects of aggregation.	33
2.3	The relationship between MNI and NISP, Prolonged Drift, Kenya.	53
2.4	The relationship between $\log_{10}$ MNI and $\log_{10}$ NISP, Prolonged Drift, Kenya.	54
2.5	The relationship between $\log_{10}$ MNI and $\log_{10}$ NISP, Apple Creek site, Illinois.	55
2.6	The relationship between $\log_{10}$ MNI and $\log_{10}$ NISP, Buffalo site, West Virginia.	56
2.7	The relationship between $\log_{10}$ MNI and $\log_{10}$ NISP, Dirty Shame Rockshelter, Oregon, Stratum 2.	57
2.8	The relationship between $\log_{10}$ MNI and $\log_{10}$ NISP, Dirty Shame Rockshelter, Oregon, Stratum 4.	58
2.9	The relationship between $\log_{10}$ MNI and $\log_{10}$ NISP, Fort Ligonier, Pennsylvania.	59
2.10	The relationship between $\log_{10}$ MNI and $\log_{10}$ NISP, bird remains from Arikara village sites.	60
2.11	The relationship between MNI and NISP for five species of shrews, Clark's Cave, Virginia.	61

LIST OF FIGURES

2.12	The relationship between $\log_{10}$ NISP and $\log_{10}$ MNI for the Dirty Shame Rockshelter Stratum 4 mammals.	62
2.13	The relationship between $\log_{10}$ MNI$_{10cm}$ and $\log_{10}$ NISP for the Connley Cave No. 4 mammals.	64
2.14	The relationship between $\log_{10}$ MNI$_{stratum}$ and $\log_{10}$ NISP for the Connley Cave No. 4 mammals.	65
2.15	The limits between which the relationship of MNI/NISP to NISP must vary.	69
2.16	The relationship between $\log_{10}$ MNI/NISP and $\log_{10}$ NISP for the Buffalo site birds and mammals.	70
2.17	The relationship between $\log_{10}$ NISP/MNI and $\log_{10}$ NISP for the Prolonged Drift mammals.	71
2.18	The relationship between $\log_{10}$ NISP/MNI and $\log_{10}$ NISP for the Buffalo site birds and mammals.	74
2.19	The relationship between $\log_{10}$ MNI/NISP and $\log_{10}$ NISP for the Prolonged Drift mammals.	76
2.20	The relationship between $\log_{10}$ CSI and $\log_{10}$ NISP for the Boardman fauna.	78
2.21	The relationship between $\log_{10}$ CSI and $\log_{10}$ NISP for the Hemphill fauna.	79
2.22	The relationship between Thomas' coefficient B and $\log_e$ NISP for the Hanging Rock Shelter mammals.	82
2.23	The relationship between $\log_{10}$ MNI/NISP and $\log_{10}$ NISP for the Connley Cave No. 4 mammals: MNI$_{10cm}$.	84
2.24	The relationship between $\log_{10}$ MNI/NISP and $\log_{10}$ NISP for the Connley Cave No. 4 mammals: MNI$_{stratum}$.	85
2.25	The relationship between $\log_{10}$ MNI/NISP and $\log_{10}$ NISP for the Connley Cave No. 4 mammals: MNI$_{site}$.	86
3.1	Minimum numbers, numbers of identified specimens, and "actual" abundances of three taxa in a hypothetical fauna.	95
3.2	The distribution of taxonomic abundances for the Apple Creek midden and plow-zone mammals.	97
3.3	The distribution of taxonomic abundances for the Buffalo site mammals.	98
3.4	The distributions of taxonomic abundances for the Dirty Shame Rockshelter Stratum 2 mammals.	99
3.5	The distribution of taxonomic abundances for the Dirty Shame Rockshelter Stratum 4 mammals.	100
3.6	The distribution of taxonomic abundances for the Fort Ligonier mammals.	101
3.7	The distribution of taxonomic abundances for the Prolonged Drift mammals.	102

LIST OF FIGURES

3.8	The relationship between $\log_{10}$ MNI and $\log_{10}$ NISP for the five most abundant taxa at Prolonged Drift.	111
4.1	The distribution of the remains of deer and of all other vertebrates by level within Raddatz Rockshelter.	129
5.1	Taxonomic frequency structure of a faunal assemblage in which all identified specimens belong to the same taxon.	132
5.2	Taxonomic frequency structure of a faunal assemblage in which all identified specimens belong to different taxa.	133
5.3	The relationship between numbers of identified specimens and numbers of taxa.	134
5.4	The distribution of taxonomic abundances for the Hidden Cave Stratum IV mammals.	135
5.5	The distribution of taxonomic abundances within a 33% random sample of the Hidden Cave Stratum IV mammal specimens identified to the species level.	137
5.6	The relationship between assemblage richness ($\log_{10}$ number of species per assemblage) and $\log_{10}$ NISP across all Gatecliff Shelter small-mammal assemblages that provided at least one specimen identified to the species level.	138
5.7	The relationship between assemblage generic richness ($\log_{10}$ number of genera per assemblage) and $\log_{10}$ NISP across all Gatecliff Shelter small-mammal assemblages that provided at least one specimen identified to the genus level.	140
5.8	The relationship between $\log_{10}$ number of species per assemblage and $\log_{10}$ NISP across all Hidden Cave mammalian assemblages that provided at least one specimen identified to the species level.	141
5.9	The relationship between $\log_{10}$ number of genera per assemblage and $\log_{10}$ NISP across all Hidden Cave mammalian assemblages that provided at least one specimen identified to the genus level.	143
5.10	The relationship between $\log_{10}$ number of species per assemblage and $\log_{10}$ NISP across all Meadowcroft Rockshelter mammalian assemblages.	144
5.11	The relationship between $\log_{10}$ number of species per assemblage and $\log_{10}$ NISP across Meadowcroft Rockshelter mammalian assemblages except Stratum VIII.	145
5.12	The relationship between $\log_{10}$ number of species and $\log_{10}$ NISP across 11 Fremont avian assemblages.	147
5.13	The relationship between $\log_{10}$ number of species and $\log_{10}$ NISP across 17 Fremont mammalian assemblages.	148
5.14	The relationship between $\log_{10}$ number of species (Y) and	149

	$\log_{10}$ NISP across 17 Fremont mammalian assemblages: plot of residuals in unit normal deviate form against predicted Y values.	
5.15	The relationship between number of species and $\log_{10}$ NISP across 17 Fremont mammalian assemblages.	150
5.16	The relationship between number of species (Y) and $\log_{10}$ NISP across 17 Fremont mammalian assemblages: plot of residuals in unit normal deviate form against predicted Y values.	151
6.1	Percentage of identified specimens recovered per body-size class for Thomas' three Nevada faunas: ¼-inch (.64 cm) screen.	170
6.2	Cumulative percentage of identified specimens recovered by screen size for Thomas' three Nevada faunas: Class I mammals.	171
6.3	Number of Common Goldeneyes present by week during 1965 on the Lower Klamath National Wildlife Refuge, northeastern California.	175
6.4	Number of Redheads present by week during 1965 on the Lower Klamath National Wildlife Refuge, northeastern California.	176

List of Tables

1.1	Number of identified specimens per small mammal taxon by stratum at Gatecliff Shelter.	6
1.2	Chronology of the Hidden Cave deposits.	12
1.3	Number of identified specimens per mammal taxon by stratum at Hidden Cave.	13
2.1	An example of the effects of bone fragmentation on the statistical assessment of significant differences between faunas.	23
2.2	The effects of aggregation on minimum numbers: abundance ratios unaltered.	31
2.3	The effects of aggregation on minimum numbers: abundance ratios altered.	32
2.4	Numbers of identified specimens by stratum, Connley Cave No. 4 mammals.	35
2.5	Total minimum numbers of individuals by aggregation method, Connley Cave No. 4 mammals.	37
2.6	Maximum possible differences in minimum number values, Connley Cave No. 4 mammals.	38
2.7	Distribution of most abundant elements within Stratum 3 for two Connley Cave No. 4 mammals.	39
2.8	The effects of aggregation on abundance ratios based on minimum numbers: selected pairs of Connley Cave No. 4 mammals.	39
2.9	Comparing taxonomic abundances between strata by analytic approach: Connley Cave No. 4 *Lepus* spp.	41

2.10	Minimum numbers of individuals for the Hidden Cave mammals, calculated by separate and by grouped strata.	42
2.11	Comparisons of relative abundance of *Sylvilagus* and *Lepus* in Strata I–V and VI–X, Hidden Cave.	45
2.12	The calculation of percentage survival of skeletal parts: an example.	46
2.13	The calculation of percentage survival of skeletal parts: Hidden Cave *Lepus*, Strata I–V, using minimum numbers calculated on a single stratum basis.	47
2.14	The effects of aggregation on the calculation of percentage survival of skeletal parts: Hidden Cave *Lepus*, Strata I–V.	48
2.15	Minimum numbers of individuals and numbers of identified specimens per taxon, Prolonged Drift, Kenya.	52
2.16	Regression equations and correlation coefficients for the relationship between MNI and NISP at Prolonged Drift, Apple Creek, Buffalo, Dirty Shame Rockshelter Stratum 2, Dirty Shame Rockshelter Stratum 4, and Fort Ligonier.	54
2.17	Identified specimens of shrews from Clark's Cave, Virginia.	60
2.18	Regression equations and correlation coefficients for the relationship between NISP and MNI at Prolonged Drift, Apple Creek, Buffalo, Dirty Shame Rockshelter Stratum 2, Dirty Shame Rockshelter Stratum 4, and Fort Ligonier.	63
2.19	Regression equations and correlation coefficients for the relationship between NISP and MNI_{10cm} and $MNI_{stratum}$ for the Connley Cave No. 4 mammals.	65
2.20	Regression equations and correlation coefficients for the relationship between MNI/NISP and NISP at Prolonged Drift, Apple Creek, Buffalo, Dirty Shame Rockshelter Stratum 2, Dirty Shame Rockshelter Stratum 4, and Fort Ligonier.	72
2.21	Numbers of identified specimens, minimum numbers of individuals, and NISP/MNI for deer and woodchuck at the Buffalo site.	73
2.22	Regression equations and correlation coefficients for the relationship between CSI and NISP for the Boardman, Hemphill, McKay, and Westend Blowout faunas.	78
2.23	Average number of identified specimens per taxon within Shotwell's proximal and more distant communities: Boardman, Hemphill, McKay, and Westend Blowout.	80
2.24	Regression equations and correlation coefficients for the relationship between Thomas' coefficient B and NISP at Hanging Rock Shelter, Little Smoky Creek Shelter, and Smoky Creek Cave.	82
2.25	Regression equations and correlation coefficients for the rela-	83

LIST OF TABLES

	tionship between MNI/NISP and NISP for the Connley Cave No. 4 mammals.	
3.1	Ratios of taxonomic abundance for the taxa illustrated in Figure 3.1.	95
3.2	Rank orders of abundance from all abundance measures: Connley Cave No. 4 mammals.	103
3.3	Rank order of rearrangements of ordinal taxonomic abundances for all pairs of abundances measures, Connley Cave No. 4 mammals.	103
3.4	The five most abundant mammalian taxa at Prolonged Drift.	104
3.5	The ten most abundant mammalian taxa at the Buffalo site.	104
3.6	The four most abundant taxa at Fort Ligonier.	105
3.7	NISP- and MNI-based rank order abundances for the seven most abundant mammalian taxa at Hidden Cave.	108
3.8	Minimum numbers of individuals (MNI_{10cm}) and numbers of identified specimens (NISP) of pika and all other mammals by stratum within Connley Caves 3, 4, 5, and 6.	113
3.9	Numbers of identified pika specimens by grouped strata within Gatecliff Shelter.	114
4.1	Minimum numbers of individuals and relative abundances of deer by phase at Snaketown.	118
4.2	Sample sizes and relative abundances of xeric rodents by major stratigraphic unit, Hogup Cave.	119
4.3	Sample sizes and relative abundances of deer, Raddatz Rockshelter.	120
4.4	Relative abundances of *Sylvilagus idahoensis* through time at Gatecliff Shelter: all strata.	122
4.5	Relative abundances of *Sylvilagus idahoensis* at Gatecliff Shelter: strata with more than 150 total identified specimens.	123
4.6	Relative abundances of *Lepus* through time at Gatecliff Shelter: all strata.	124
4.7	Relative abundances of *Lepus* through time and Gatecliff Shelter: strata with more than 150 identified specimens.	125
4.8	Relative abundances of *Marmota flaviventris* through time at Hidden Cave: all strata with identified mammalian specimens.	126
4.9	Relative abundances of *Marmota flaviventris* through time at Hidden Cave: strata with 145 or more identified mammalian specimens.	127
4.10	The number of identified deer bones and of all other vertebrates by level at Raddatz Rockshelter.	128
5.1	Hidden Cave Stratum IV species NISP, and a 33% random sample of species NISP.	136
5.2	Numbers of specimens identified to the species and genus	139

	levels, numbers of species, and numbers of genera: Gatecliff Shelter small mammals.	
5.3	Numbers of specimens identified to the species and genus levels, number of species, and numbers of genera: Hidden Cave mammals from unmixed strata.	142
5.4	Numbers of specimens identified to the species level and numbers of species: Meadowcroft Rockshelter mammals.	142
5.5	Species richness and sample size: regression equations and correlation coefficients for the Gatecliff, Hidden Cave, and Meadowcroft mammals.	144
5.6	Numbers of specimens identified to the species level and numbers of species for 11 Fremont avian assemblages.	146
5.7	Numbers of specimens identified to the species level and numbers of species for 17 Fremont mammalian assemblages.	146
5.8	Assemblage species richness and sample size: regression equations and correlation coefficients for 11 avian and 17 mammalian faunal assemblages from Fremont sites.	150
5.9	Numbers of identified specimens per mammalian species at the Bear River 1 and Bear River 3 sites.	154
5.10	Adjusted residuals for the Bear River 1 and Bear River 3 mammalian species assemblages.	155
5.11	Calculation of the Smirnov test statistic for the Bear River 1 and Bear River 3 mammalian species assemblages.	156
5.12	Smirnov test statistic T' for a series of Fremont mammalian species assemblages.	157
5.13	$NISP_i/\Sigma$ NISP for gopher tortoise and shark and total numbers of identified specimens per level, Jungerman fauna.	160
5.14	Diversity indices and sample sizes, Jungerman fauna.	161
5.15	Diversity values and total numbers of identified specimens for 13 Fremont mammalian assemblages.	162
5.16	Diversity values and total numbers of identified specimens for nine Fremont avian species assemblages.	162
5.17	Diversity values and total minimum numbers of individuals for nine Fremont avian species assemblages.	163
5.18	Numbers of identified specimens for the most abundant taxon, total numbers of identified specimens, and $NISP_i/\Sigma$ NISP for nine Fremont avian species assemblages.	163
5.19	Numbers of identified specimens for the most abundant taxon, total numbers of identified specimens, and $NISP_i/\Sigma$ NISP for 13 Fremont mammalian species assemblages.	164
5.20	Minimum numbers of individuals for the most abundant taxon, total minimum numbers of individuals, and MNI_i/Σ MNI for nine Fremont avian species assemblages.	165

LIST OF TABLES

5.21	Specimen-based diversity values for Fremont avian species assemblages from marsh and non-marsh environmental settings.	167
6.1	Numbers of identified specimens collected by screen size and by body-size class from three Nevada sites.	169
6.2	Percentage recovery by screen size and body-size class for Thomas' three Nevada faunas.	170

Preface

It used to be easy to study vertebrate faunal remains, and in particular bones and teeth, from archaeological sites: identify the bones, add up the numbers, write the report. In the last few years that situation has changed dramatically. Detailed studies have been made of the processes that transform living animals into the bones and teeth that archaeologists and paleontologists retrieve from the ground. These studies have demonstrated that the transforming processes are often so complex that it is difficult to know what it is faunal analysts are measuring when they add up their numbers. As if that were not enough, detailed studies of the "behavior" of counting units faunal analysts use have suggested that these units have quirks and oddities that are far from benign and that are far from fully understood. Although many faunal analysts still simply identify the bones, add up the numbers, and write the report, their numbers are rapidly diminishing. They are being replaced by analysts who give much thought to taphonomic processes and to problems of quantification.

This book is about problems in the quantification of bones and teeth from archaeological and, to some extent, paleontological sites. In it, I deal with the units that are routinely used to measure the abundances of the animals that contributed their bones and teeth to a given set of faunal remains, and with various kinds of statistical manipulations that are or can be done with those measurements. My goal is to help make known the various quirks and oddities that characterize those measurements, and to present ways in which faunal analysts can at least detect, and perhaps even avoid, the pitfalls that those measurements seem to present at every turn.

I have written this book primarily for archaeological faunal analysts, be they of archaeological, zoological, or interdisciplinary background. I believe, however, that much of what I argue has relevance to paleontological faunas as well. Indeed, the general kinds of problems that I discuss in the chapters that follow affect a wide variety of quantitative archaeological studies, including a surprisingly large number of lithic and ceramic analyses. As a result, it is my hope that archaeologists in general will benefit from this volume.

My sincere thanks go to R. Lee Lyman, Nancy Sharp, Jan Simek, and David Hurst Thomas for their critical comments on an early version of this manuscript, and especially to R. Lee Lyman for his very sharp pencil. I also thank C. Melvin Aikens for providing me with hard-to-find materials relating to Fremont, Robert C. Dunnell for sharing his thoughts on a number of the topics discussed in this volume, Patricia Ruppé for her sharp eyes, Robert D. Leonard for the help he has given me over the years, and Bonnie Whatley Styles for providing me with unpublished manuscripts relating to her work. Over the years, Richard W. Casteel and I spent innumerable hours discussing many of the issues treated here, to the point that it is at times hard to tell who thought of what first (although the approach I take in my treatment of the effects of collection techniques is borrowed entirely from his work). While we remain close friends, I have very much missed this interchange since his retirement from archaeology, and I thank him for his collegiality. My thanks also to David Hurst Thomas, who understands the meaning of interdisciplinary research; to Paul W. Parmalee, whose excellence as a faunal analyst is matched only by his remarkable patience; and to Karl W. Butzer, who accepted this volume during his tenure as editor of the Studies in Archaeological Science.

CHAPTER 1

Introduction

A decade ago, a student interested in learning all that had been written on the methodological aspects of the analysis of vertebrate faunal remains from archaeological sites could satisfy that interest by spending a few days reading in the library. Even simple issues, such as the effects of different approaches to clustering faunal material on the definition of the minimum number of individuals (Grayson 1973), went virtually unexplored. Although there were some important exceptions, much of the literature that seemed of greatest methodological value to the analysis of archaeological faunas had been produced by paleontologists, dealt with paleontological problems, and had been published in paleontological journals and monographs. Even the paper most frequently cited by American archaeological faunal analysts had been written by a paleontologist, Theodore E. White, for archaeological consumption (White 1953).

Today, the situation is remarkably different. It now takes longer to type the list of methodologically important publications dealing with archaeological vertebrate faunal analysis than it took to read the crucial material in 1970. Although there is a long tradition of specialization in archaeozoology in Europe, there were few such specialists in North America a decade ago. One would begin struggling for names after mentioning John Guilday and Paul Parmalee. Today, there are many North American specialists with strong interdisciplinary backgrounds, most of whom have made worthwhile contributions to the methodological literature. At the same time, the number of methodological contributions by Europeans and others has grown rapidly. Since 1980, four books on vertebrate taphonomy alone have appeared (Behrensmeyer and Hill 1980; Binford 1981; Brain 1981; Shipman 1981), and the literature on the quantification of archaeological vertebrates is accumulating quickly as well. There is now much more than can be read in a few days.

The zooarchaeological literature has grown in diverse ways, however. Analysts interested in quantification must take into account the vagaries introduced by taphonomic factors. After all, taphonomic considerations were crucial in generating concern over the quantification of archaeological and paleontological faunas in the first place. Rarely can one find a paper on quantification that does not also discuss problems introduced by the complexities of taphonomic pathways. On the other hand, taphonomists have tended to skirt issues raised by those interested in such problems as the quantification of taxonomic abundances within vertebrate faunas. That this is so is clear not only to analysts who have focused on counting issues, but also to observers outside the realm of faunal analysis itself (e.g., Dunnell 1982).

Certainly, not all taphonomists ignore such problems (e.g., Gifford 1981), and many taphonomic studies do not deal with issues of quantification simply because their focus is on processes, not on quantifying the results of those processes as they appear to archaeologists and paleontologists thousands of years later (e.g., Behrensmeyer 1978; Behrensmeyer and Boaz 1980). Some taphonomists may also skirt issues of quantification because we know so little about taphonomic processes that it is not clear that the detailed numerical structures of faunas that have been produced by those processes are to be trusted (see, for instance, Voorhies' 1969 criticism of the methods of paleoecological reconstruction proposed by Shotwell 1955, 1958, 1963). But there are also taphonomic studies that combine detailed considerations of processes with detailed analyses of the numerical structure of the faunas produced by those processes. Here, lack of concern with the particular problems posed by the quantification of vertebrate faunas can be extremely harmful. Brain (1981) provides an example. Brain's book is a fascinating contribution filled with insightful considerations of the processes that convert living animals into the shreds of bone and tooth that archaeologists and paleontologists recover. It is also filled with statistical manipulations of faunal counts without appreciation for the difficulties associated with such manipulations. In a work of this sort, skirting issues of quantification can be dangerous.

My focus in this book is on a series of selected issues concerning the quantification of vertebrate faunas from archaeological and, to some extent, paleontological sites. It will become evident that I have been very selective in my choice of topics. I focus primarily on matters that relate to the measurement of taxonomic abundances in archaeological and paleontological faunas, and to the manipulation of those abundances. Throughout, I am concerned with the interrelationship between various abundance measures — from simple counts of relative abundances to the measurement of faunal diversity — and the size of the samples on which those measures are based. I am not a taphonomist, and deal with taphonomy only insofar as our current knowledge of taphonomic processes defines and limits the kinds of inferences that can validly be made from faunas excavated from the ground or collected from its surface. My purpose in writing this book is to tie together, and in a number of cases to modify, the arguments that I have previously made concerning the hazards involved in quantifying vertebrate faunas. I hope to show the importance of considering such hazards whenever faunal abundances are measured. In nearly every way, this volume represents my own, perhaps idiosyncratic, views on the topics I cover, and I emphasize that I have made no attempt to review or to take issue with the literature that has dealt with these topics in other ways. This is not, in short, a textbook on the quantification of archaeological and paleontological vertebrate faunas. Some of the arguments I present here have been published in articles during the past decade. With few exceptions, details, and in some cases even outcomes, of those arguments have changed as I have continued to consider the issues involved. Finally, I note that whether or not I am correct about the solutions I propose to the problems discussed here, the problems are real and must be dealt with by anyone who attempts to measure the abundances of taxa within archaeological or paleontological faunas, and to analyze those measures.

Background to Two Faunal Samples

Although my discussion draws on many sites from many times and places, I have chosen to illustrate most of my analyses with two faunas from western North American rockshelters. The reasons for illustrating a diverse set of issues in faunal quantification with the same set of faunas (as well as with a diverse variety of others) are simple: there is value in the continuity provided by demonstrating very different problems with the same set or sets of faunal materials, in demonstrating why solutions to some kinds of problems work in different faunas, and in demonstrating why solutions that work in one case do not necessarily work in another. The two faunas I have selected come from sites located in the state of Nevada: Gatecliff Shelter and Hidden Cave. I chose these sites for several reasons. First, I conducted the original analyses of both faunas and thus am more familiar with them than I am with the examples I draw from the published works of others. Second, both have reasonably large samples but differ greatly in the size of those samples and in the number of taxa represented. Third, both come from settings that have been studied in great depth: there is detailed information available on the archaeology, stratigraphy, chronology, and paleobotany of each. Finally, the deposits of both sites are stratified and span long periods of time, attributes that will be of importance to my analyses. Since I use one or both of these faunas in nearly every chapter, I have provided basic information on them here.

Gatecliff Shelter

Gatecliff Shelter is located in the Toquima Range of central Nevada, at an elevation of 2320 m. Today, the site is surrounded by vegetation characterized by big sagebrush (*Artemisia tridentata*), green rabbitbrush (*Chrysothamnus viscidiflorus*), Mormon tea (*Ephedra viridis*), single-needle piñon (*Pinus monophylla*), Utah juniper (*Juniperus osteosperma*), and scattered grasses (see the discussion in Thompson 1983). A few hundred feet south of the site runs Mill Creek, the small stream that drains the canyon in which Gatecliff Shelter is located.

Excavated by David H. Thomas of the American Museum of Natural History, Gatecliff Shelter contained a remarkably well-stratified record of human occupation extending back about 7000 years. Approximately 600 m^3 of fill was removed from Gatecliff, the excavations reaching a depth of over 10 m. Besides Thomas himself, a wide variety of specialists played a role in the analysis of the materials retrieved from the site, including stratigraphers, paleobotanists, and myself (see Thomas 1983a,b).

The stratigraphy of Gatecliff is remarkable. Jonathan O. Davis and his colleagues defined a series of 56 geological strata within the shelter, some of which contain several discrete cultural horizons. The earliest of these strata dates to some time before, but not long before, 7000 B.P. With the exception of the three lowest strata, all of which were composed of rock rubble, the stratigraphic sequence for Gatecliff consists of a series of layers of rocky rubble separated by graded alluvial beds ranging from a few centimeters to nearly a meter in thickness. The layers of rocky rubble were deposited as a result of debris flows from the adjacent slopes of Mill Canyon, talus activity, and roof fall. The graded alluvial beds appear to represent the ultimate product of debris flows that originated upstream from the site. These flows created surges of sediment-charged waters that periodically entered Gatecliff. The debris-flow gravels and graded alluvial beds represent brief periods of time; only the slowly accumulating roof fall and talus rubbles, and the surfaces provided by the debris-flow rubbles and alluvial beds, record the passage of lengthy periods. As a result, the *in situ* record of human occupation at Gatecliff is associated with the talus rubbles and stable surfaces. Materials found in the debris-flow gravels and alluvial beds would have been derived from elsewhere in Mill Canyon, or have infiltrated downwards from strata above (Davis *et al.* 1983). In fact, however, the graded alluvial beds contain neither artifacts nor bones.

The chronology of the Gatecliff Shelter deposits is based primarily on radiocarbon determinations, although additional information is available from obsidian hydration dates. Further, Stratum 55 contains Mazama tephra, well dated to ca. 6900 B.P. The single rubble stratum beneath Stratum 55 is undated and contains no artifacts, but there is no reason to think that it predates the eruption of Mt. Mazama by a significant amount of time. The precise dating of the Gatecliff strata will be presented in later chapters as needed.

Gatecliff was excavated using $\frac{1}{8}$-inch (.32-cm) mesh screen. Although there is reason to believe that small specimens were lost through even this fine a mesh (see Chapter 6), the care taken in excavating the Gatecliff deposits, and the fact that these deposits were rich in vertebrate remains, resulted in the retrieval of a large collection of mammalian bones and teeth, nearly all of which came from

BACKGROUND TO TWO FAUNAL SAMPLES 5

extremely well-controlled stratigraphic settings. Approximately 14,000 mammalian specimens were identified to at least the genus level from these deposits. Of these, some 13,000 represent small mammals, and are the focus of attention in later chapters (Grayson 1983b; Thomas 1983b; see Table 1.1).

Hidden Cave

Well within the drainage of Pleistocene Lake Lahontan, Hidden Cave is located on the northern edge of Eetza Mountain, an outlier of western Nevada's Stillwater Range, at an elevation of 1251 m. Today, the vegetation surrounding the site is dominated by desert shrubs: little greasewood (*Sarcobatus baileyi*), saltbush (*Atriplex confertifolia*), and, much less commonly, budsage (*Artemisia spinescens*). Looking south from the mouth of the cave, the view is dominated by playas and sand dunes, with big greasewood (*S. vermiculatus*) dominating the vegetation. The modern fauna in the immediate vicinity of Eetza Mountain consists of species well-adapted to hot and dry summer conditions (Grayson 1984; Kelly and Hattori 1984).

As the name suggests, Hidden Cave is a true cave. Formed some 21,000 years ago when the waves of Pleistocene Lake Lahontan eroded a cavity in the side of Eetza Mountain, Hidden Cave has a maximum length of 45 m and a maximum width of 29 m. Prior to excavation, the maximum distance from cave floor to cave roof was 4.5 m. Until recently, entry was through a small, barely visible portal just large enough to admit a person in full crouch. Before the 1920s, the cave was even better hidden; erosion at this time revealed a small opening leading into Eetza Mountain, an opening that was enlarged by the boys who discovered it (Thomas 1984).

The cave has been excavated a number of times during the past five decades. The Nevada State Park Commission excavated here in 1940, but the results of this work were never published. Further work was conducted in the cave in 1951 by the University of California, Berkeley. While the full results of this excavation also went unpublished, the geologist Roger B. Morrison had taken part in the work, and subsequently published a stratigraphic section and analysis of the Hidden Cave deposits as part of his study of the history of Lake Lahontan (Morrison 1964). Hidden Cave thus became a geologically important site, having provided a stratigraphic type section for the later history of the southern Lahontan Basin. In addition to describing the stratigraphy of the Hidden Cave deposits, Morrison (1964) provided lists of the vertebrate species represented by the fossils and subfossils retrieved by the 1951 excavations. Most of these specimens were of mammals, identified by the paleontologists S. B. Benson and E. L. Furlong; a smaller series of birds were identified by the paleontologist Hildegarde Howard. The zoological data provided by Morrison (1964) represented the first deep, stratified sequence of vertebrate faunal remains published for any part of the Lahontan Basin.

Table 1.1

Number of Identified Specimens per Small Mammal Taxon by Stratum at Gatecliff Shelter

	Sorex vagrans	Ochotona princeps	Sylvilagus sp.	Sylvilagus cf. idahoensis	Sylvilagus idahoensis	Sylvilagus cf. nuttallii	Sylvilagus nuttallii	Lepus sp.
Stratum 1								
Horizon 1	—	—	34	—	11	96	13	18
Horizon 2	1	—	139	—	35	419	50	50
Horizon 3	—	2	168	—	39	727	97	45
Horizons 1/3	—	—	7	—	1	36	6	5
Stratum 2	—	—	16	—	4	60	4	6
Stratum 3–5								
Horizon 4	1	—	120	—	20	441	67	35
Horizon 5	—	—	70	—	14	307	37	17
Horizon 6	—	—	146	—	18	535	82	36
Horizons 4/6	—	—	51	—	11	229	27	26
Stratum 6/7	—	—	31	—	6	120	22	4
Stratum 8	—	—	13	—	7	28	1	2
Stratum 9	—	1	64	—	10	278	50	31
Stratum 10	—	—	—	—	—	5	1	17
Stratum 11/12	—	—	62	—	23	189	21	—
Stratum 13	—	—	13	—	3	15	3	1
Stratum 14–16	—	—	—	—	—	1	—	—
Stratum 17	—	—	3	—	—	13	—	1
Stratum 18	—	—	—	—	—	—	—	—
Stratum 19	—	—	19	—	7	41	9	4
Stratum 20	—	1	2	—	2	14	1	1
Stratum 21	—	—	1	—	—	—	—	—
Stratum 22	—	1	8	—	8	19	1	10
Stratum 23	—	—	1	—	—	9	—	5
Stratum 24/25	—	11	24	—	47	45	1	32
Stratum 26–30	—	—	3	—	—	8	—	—
Stratum 31/32	—	5	20	—	35	36	2	24
Stratum 33	—	3	53	—	41	124	8	27
Stratum 37	—	9	18	—	63	46	—	36
Stratum 54	—	7	40	1	68	83	4	62
Stratum 55	—	—	—	—	—	—	—	—
Stratum 56	—	18	86	5	263	137	10	170
Mixed strata:								
Strata 1–5	—	—	1	—	—	9	1	—
Strata 3–7	—	—	14	—	—	33	6	5
Strata 3–8	—	—	—	—	—	2	—	—
Totals	2	58	1227	6	736	4105	524	670

BACKGROUND TO TWO FAUNAL SAMPLES

Table 1.1 (*Continued*)

Lepus cf. *townsendii*	*Lepus* cf. *californicus*	*Eutamias* sp.	*Eutamias* cf. *minimus*	*Eutamias* *minimus*	*Eutamias* cf. *dorsalis*	*Eutamias* *dorsalis*	*Eutamias* cf. *umbrinus*	*Eutamias* *umbrinus*
—	2	—	—	—	1	—	1	—
—	7	7	1	2	2	—	—	—
2	5	11	8	1	5	1	2	—
—	—	—	—	—	1	—	—	—
—	—	—	—	—	—	—	—	—
—	4	8	3	1	2	—	1	—
—	1	10	1	—	—	1	2	—
—	1	15	9	4	2	—	—	—
—	1	1	—	7	3	—	—	—
—	—	2	—	—	—	—	—	—
—	—	1	—	—	—	—	—	—
—	1	3	4	1	—	2	1	—
—	—	—	1	—	—	—	—	—
—	—	3	7	5	1	—	1	—
—	—	62	8	4	3	4	1	—
—	—	—	—	—	—	—	—	—
—	—	9	—	—	—	13	—	—
—	—	1	—	—	—	—	—	—
—	—	37	—	—	1	3	—	—
—	2	1	—	—	—	—	—	—
—	—	—	—	—	—	—	—	—
—	—	3	—	—	—	—	—	—
—	—	—	—	—	—	—	—	—
—	1	9	—	1	1	—	—	—
—	—	31	—	—	—	—	—	1
—	1	—	—	—	—	—	—	—
—	1	1	1	—	—	—	—	—
—	1	1	—	—	—	—	—	—
—	5	3	—	1	—	—	—	—
—	—	—	—	—	—	—	—	—
—	16	7	1	—	—	6	—	—
—	—	1	—	—	—	—	—	—
—	—	2	—	—	—	—	—	—
—	—	—	—	—	—	—	—	—
2	49	229	44	27	22	30	9	1

Table 1.1 (*Continued*)

	Marmota flaviventris	*Ammospermophilus leucurus*	*Spermophilus* sp.	*Spermophilus townsendii*	*Spermophilus beldingi*	*Spermophilus lateralis*	*Thomomys* sp.	*Thomomys* cf. *bottae*	*Thomomys bottae*
Stratum 1									
Horizon 1	1	—	6	4	—	2	7	1	—
Horizon 2	1	—	26	23	—	7	16	—	—
Horizon 3	1	—	34	10	—	14	24	—	2
Horizons 1/3	—	—	2	1	—	1	2	1	2
Stratum 2	1	—	4	3	—	1	3	—	1
Stratum 3–5									
Horizon 4	1	2	28	8	—	6	16	1	—
Horizon 5	—	—	25	12	3	8	4	—	1
Horizon 6	—	—	26	9	—	14	24	1	—
Horizons 4/6	—	—	8	3	—	7	8	1	—
Stratum 6/7	—	—	2	5	—	3	—	—	—
Stratum 8	—	—	—	4	—	—	—	—	—
Stratum 9	—	1	4	2	—	2	12	—	—
Stratum 10	—	—	—	—	—	—	—	—	—
Stratum 11/12	—	—	5	2	—	3	12	—	—
Stratum 13	—	—	5	2	—	1	6	—	—
Stratum 14–16	—	—	—	—	—	—	—	—	—
Stratum 17	—	—	1	1	—	—	—	—	—
Stratum 18	—	—	1	—	—	—	—	—	—
Stratum 19	—	—	7	1	—	1	9	1	1
Stratum 20	—	—	2	1	—	—	3	—	1
Stratum 21	—	—	—	—	—	—	—	—	—
Stratum 22	—	—	3	1	—	—	5	—	—
Stratum 23	—	—	—	1	—	—	1	—	—
Stratum 24/25	—	—	84	30	—	—	25	—	1
Stratum 26–30	—	—	3	1	—	—	—	—	—
Stratum 31/32	—	—	2	4	—	—	20	2	1
Stratum 33	—	—	—	—	—	—	7	—	—
Stratum 37	—	—	—	—	—	—	1	—	—
Stratum 54	1	—	5	—	—	—	4	1	—
Stratum 55	—	—	—	—	—	—	15	—	—
Stratum 56	—	—	4	2	—	—	53	1	—
Mixed strata:									
Strata 1–5	1	—	—	—	—	—	—	—	—
Strata 3–7	—	—	—	2	—	1	—	—	—
Strata 3–8	—	—	—	—	—	—	—	—	—
Totals	7	3	287	132	3	71	277	10	10

BACKGROUND TO TWO FAUNAL SAMPLES

Table 1.1 (*Continued*)

Perognathus sp.	*Perognathus parvus*	*Microdipodops megacephalus*	*Dipodomys* sp.	*Peromyscus* sp. subgenus?	*Peromyscus* sp. subgenus *Peromyscus*	*Peromyscus crinitus*	*Onychomys* sp.	*Neotoma* sp.
9	3	—	—	3	4	—	—	12
19	6	—	5	7	—	1	—	46
40	17	—	2	19	6	2	—	94
1	—	—	1	1	1	—	—	—
1	—	—	2	1	1	—	1	6
30	17	—	3	4	2	4	—	33
9	3	—	—	2	4	—	—	25
13	5	1	3	6	4	—	—	28
17	3	—	—	8	1	—	—	5
3	—	—	—	—	1	—	—	8
—	—	—	1	1	1	1	—	3
5	1	—	—	2	3	—	—	20
—	—	—	—	—	—	—	—	1
8	—	—	1	3	7	1	—	30
10	3	—	2	1	2	—	—	10
—	—	—	—	—	—	—	—	—
2	1	—	—	—	—	—	—	4
—	—	—	—	—	—	—	—	—
27	7	—	—	9	4	—	—	13
1	—	—	1	—	1	—	—	—
1	—	—	—	—	—	—	—	—
—	1	—	—	1	—	1	—	1
—	1	—	1	1	—	—	—	2
—	—	—	—	4	—	—	1	17
—	—	—	1	—	2	—	—	2
4	—	—	—	1	—	—	—	7
3	—	—	1	7	2	1	—	23
1	—	—	—	—	—	—	1	5
1	—	—	3	1	1	—	—	20
—	—	—	—	—	—	—	—	—
2	—	—	—	4	2	1	—	50
—	—	—	—	1	—	—	—	1
1	—	—	—	1	1	1	—	2
208	68	1	27	88	50	13	3	468

Table 1.1 (*Continued*)

	Neotoma cf. *lepida*	*Neotoma lepida*	*Neotoma* cf. *cinerea*	*Neotoma cinerea*	*Phenacomys* cf. *intermedius*	*Microtus* sp.	*Microtus* cf. *montanus*	*Microtus* cf. *longicaudus*	*Lagurus curtatus*
Stratum 1									
Horizon 1	2	12	31	41	—	6	—	—	8
Horizon 2	8	11	94	133	—	10	2	2	16
Horizon 3	14	26	165	271	—	22	1	4	44
Horizons 1/3	—	6	12	15	—	—	—	—	3
Stratum 2	1	2	17	21	—	3	—	—	2
Stratum 3–5									
Horizon 4	4	12	76	128	—	12	—	3	22
Horizon 5	1	5	34	76	—	11	—	—	9
Horizon 6	4	8	71	126	—	9	—	—	18
Horizons 4/6	—	4	19	51	—	6	—	—	6
Stratum 6/7	—	1	13	33	—	—	—	—	—
Stratum 8	—	1	9	7	—	—	—	—	1
Stratum 9	7	1	50	65	—	6	—	—	6
Stratum 10	—	—	1	2	—	1	—	—	—
Stratum 11/12	8	8	41	87	—	10	—	1	7
Stratum 13	—	5	24	43	—	2	—	—	2
Stratum 14–16	—	—	—	—	—	—	—	—	—
Stratum 17	—	—	1	4	—	—	—	—	—
Stratum 18	—	1	—	1	—	—	—	—	—
Stratum 19	—	8	38	40	—	3	—	—	1
Stratum 20	—	1	1	7	—	—	—	—	2
Stratum 21	—	—	—	—	—	—	—	—	—
Stratum 22	—	—	22	7	—	—	—	—	1
Stratum 23	—	—	6	12	—	1	—	—	—
Stratum 24/25	—	4	44	34	2	16	—	—	6
Stratum 26–30	1	2	31	17	—	2	—	—	1
Stratum 31/32	1	—	39	39	—	9	—	—	3
Stratum 33	—	1	141	108	—	12	—	—	2
Stratum 37	—	—	51	29	—	9	—	2	—
Stratum 54	3	2	87	67	—	6	—	—	11
Stratum 55	—	—	—	—	—	1	—	—	—
Stratum 56	2	—	179	142	—	22	—	—	24
Mixed strata:									
Strata 1–5	—	—	2	1	—	—	—	—	1
Strata 3–7	—	—	9	5	—	4	—	—	—
Strata 3–8	—	—	—	—	—	—	—	—	—
Totals	56	121	1308	1612	2	183	3	12	196

Table 1.1 (*Continued*)

Zapus cf. *princeps*	*Erethizon dorsatum*	*Canis* sp.	*Canis latrans*	*Vulpes vulpes*	*Spilogale gracilis*	*Mephitis mephitis*	*Lynx rufus*	Totals
—	1	—	—	—	—	—	—	329
—	—	—	—	—	2	2	2	1,152
—	4	—	1	1	—	2	2	1,935
—	—	—	—	—	—	—	—	105
—	4	—	—	—	—	—	—	165
—	8	—	—	—	2	—	1	1,126
—	6	2	—	—	—	—	1	701
—	14	—	—	—	—	—	—	1,232
—	3	—	1	—	—	—	—	507
—	12	—	—	—	—	—	—	266
—	—	—	—	—	—	—	—	81
—	2	—	1	—	—	—	—	636
—	—	—	—	—	—	—	—	29
—	—	—	—	—	—	—	—	546
—	—	—	—	—	—	—	—	235
—	—	—	—	—	—	—	—	1
—	—	—	—	—	—	—	—	53
—	—	—	—	—	—	—	—	4
—	—	—	—	—	—	—	—	291
—	—	—	—	—	—	—	—	45
—	—	—	—	—	—	—	—	2
—	1	—	—	—	—	—	—	94
—	—	—	—	—	—	—	—	41
—	—	—	—	—	—	—	—	440
—	—	—	—	—	—	—	—	106
—	—	—	—	—	—	—	—	255
1	—	—	—	—	—	—	—	568
—	—	—	—	—	—	—	—	273
—	—	—	—	—	—	—	—	487
—	—	—	—	—	—	—	—	16
—	—	—	—	—	—	—	—	1,207
—	—	—	—	—	—	—	—	19
—	5	—	—	—	—	—	—	92
—	—	—	—	—	—	—	—	2
1	60	2	3	1	4	4	6	13,041

Table 1.2
Chronology of the Hidden Cave Deposits

Stratum	Estimated age (years B.P.)
I	0-1500
hiatus	1500-3400
II	3400-3500
III	3500-3700
IV	3700-3800
hiatus	3800-5400
V	5400-6900
VI	ca. 6900
VII	6900-7500
IX	7500-10000
X	ca. 10000
XI	15000-18000
XII	ca. 18000
XIII	18000-21000
XIV	undated

The most recent excavations at Hidden Cave were conducted by David H. Thomas in 1979 and 1980. As at Gatecliff Shelter, a team of specialists participated in the project, including a stratigrapher, paleobotanists, and myself. The Hidden Cave deposits were excavated with trowel and brush, and all excavated sediments were passed through $\frac{1}{8}$-inch (.32-cm) mesh screen (Thomas and Peter 1984). Unlike Gatecliff, with its well-stratified and undisturbed deposits, the Hidden Cave sediments proved to be exceedingly complex and, in some areas, quite disturbed as a result of the activities of rodents and, at times, people. As a result, some of the items retrieved from Hidden Cave, including many bones and teeth, cannot be assigned precise stratigraphic provenience.

Davis (1984) defined 14 geological strata for the Hidden Cave deposits, strata that were often complex layers of sand, silt, and rocky rubble. These strata span the last 21,000 years. Of them, only two—Strata II and IV—contained abundant evidence of human occupation, but most strata provided floral and faunal remains. The chronology of the Hidden Cave deposits is based primarily on 15 radiocarbon dates and three identified tephra layers (two of Mazama tephra and one from the Mono-Inyo area of California). However, the fact that the deposits are quite disturbed in some areas reduces the precision of the Hidden Cave chronology (Table 1.2).

Hidden Cave provided approximately 7000 mammalian bones and teeth that were identified to at least the genus level. Of these, however, only some 3900 could be securely assigned to single strata, and it is this stratigraphically secure sample that I use in subsequent chapters (Table 1.3).

Table 1.3
Number of Identified Specimens per Mammal Taxon by Stratum at Hidden Cave

Taxon	Stratum														Totals
	I	II	III	IV	V	VI	VII	VIII	IX	X	XI	XII	XIII	XIV	
Sorex palustris	—	1	—	—	—	—	—	—	—	—	—	—	—	—	1
Myotis sp.	—	—	—	—	1	—	—	—	—	—	—	—	—	—	1
Myotis yumanensis	1	—	—	1	4	—	1	—	—	—	—	—	—	—	7
Antrozous pallidus	2	—	—	—	2	—	—	—	—	—	—	—	—	—	4
Sylvilagus sp.	3	2	1	24	21	—	12	—	7	—	—	—	2	—	72
Sylvilagus cf. nuttallii	21	70	18	221	204	—	106	—	40	—	2	—	27	1	710
Sylvilagus nuttallii	2	1	—	9	23	—	5	—	1	—	—	—	1	—	42
Lepus sp.	85	120	65	426	255	—	68	—	31	—	3	—	16	1	1070
Lepus californicus	1	1	—	3	—	—	—	—	—	—	—	—	—	—	5
Marmota flaviventris	5	10	8	140	68	—	27	—	12	—	—	—	10	—	280
Ammospermophilus cf. leucurus	—	—	—	1	—	—	1	—	—	—	—	—	1	—	3
Ammospermophilus leucurus	—	—	—	1	2	—	—	—	—	—	—	—	—	—	3
Spermophilus sp.	5	1	1	5	4	—	1	—	1	—	—	—	2	—	20
Spermophilus cf. townsendii	2	3	2	—	—	—	—	1	—	—	—	—	—	—	8
Spermophilus townsendii	2	8	5	5	4	—	1	—	—	—	—	—	—	1	26
Thomomys sp.	—	2	2	9	7	—	3	—	3	—	—	—	1	—	27
Thomomys cf. bottae	—	3	—	2	1	—	—	—	—	—	—	—	—	—	6
Thomomys bottae	—	1	—	3	1	—	1	—	—	—	—	—	—	—	6

Table 1.3 (*Continued*)

Taxon	\multicolumn{15}{c}{Stratum}
	I	II	III	IV	V	VI	VII	VIII	IX	X	XI	XII	XIII	XIV	Totals
Perognathus sp.	5	1	3	12	18	—	13	—	7	—	—	—	17	—	76
Perognathus longimembris	—	1	1	3	3	—	—	—	1	—	—	—	—	—	9
Perognathus parvus	—	—	—	2	2	—	—	—	1	—	—	—	1	—	6
Perognathus formosus	6	—	—	7	1	—	3	—	1	—	—	—	1	—	19
Microdipodops sp.	—	—	—	—	1	—	—	—	—	—	—	—	—	—	1
Dipodomys sp.	26	11	14	101	87	—	32	—	6	—	—	—	4	—	281
Dipodomys ordii	—	—	—	1	—	—	—	—	—	—	—	—	—	—	1
Dipodomys cf. *microps*	1	—	3	2	1	—	—	—	—	—	—	—	—	—	7
Dipodomys microps	5	3	3	6	7	—	2	—	—	—	—	—	—	—	26
Dipodomys cf. *deserti*	—	—	—	3	3	—	—	—	—	—	—	—	—	—	6
Dipodomys deserti	—	—	—	—	1	—	—	—	—	—	—	—	—	—	1
Reithrodontomys megalotis	—	2	—	—	—	—	—	—	—	—	—	—	—	—	2
Peromyscus sp.	—	—	—	5	5	—	3	—	—	—	—	—	2	1	16
Peromyscus maniculatus	—	1	—	8	4	—	3	—	—	—	—	—	1	—	17
Peromyscus crinitus	—	—	—	—	2	—	1	—	—	—	—	—	—	—	3
Onychomys sp.	—	—	—	—	—	—	1	—	—	—	—	—	—	—	1
Onychomys leucogaster	—	—	—	—	—	—	—	—	—	—	—	—	—	—	1
Neotoma sp.	9	12	5	55	51	—	34	—	8	—	—	—	5	—	179
Neotoma cf. *lepida*	10	7	15	51	51	—	22	—	10	—	1	—	9	1	177
Neotoma lepida	6	1	5	25	35	—	7	—	2	—	—	—	6	1	88
Neotoma cf. *cinerea*	3	6	10	95	115	—	48	—	16	—	1	—	23	1	318
Neotoma cinerea	1	2	6	47	34	—	18	—	7	—	1	—	4	—	120
Microtus sp.	4	8	15	43	22	—	6	—	3	—	—	—	—	—	101

Table 1.3 (*Continued*)

Taxon	Stratum														Totals
	I	II	III	IV	V	VI	VII	VIII	IX	X	XI	XII	XIII	XIV	
Microtus montanus	3	—	—	4	2	—	1	—	—	—	—	—	—	—	10
Ondatra zibethicus	1	—	—	4	4	—	—	—	—	—	—	—	—	—	9
Canis cf. *latrans*	—	1	—	17	3	—	—	—	—	—	—	—	—	—	21
Canis latrans	1	2	1	2	2	—	1	—	—	—	—	—	—	—	9
Canis lupus	—	1	—	1	—	—	—	—	—	—	—	—	—	—	2
Vulpes vulpes	—	—	1	6	10	—	1	—	—	—	—	—	4	—	22
Martes sp.	—	—	—	—	—	—	—	—	1	—	—	—	—	—	1
Martes nobilis	—	1	1	—	—	—	—	—	—	—	—	—	—	—	1
Mustela sp.	—	—	—	—	—	—	—	—	—	—	—	—	—	—	1
Mustela cf. *frenata*	2	—	1	3	3	—	2	—	—	—	—	—	1	—	12
Mustela frenata	—	—	—	2	1	—	1	—	—	—	—	—	—	—	4
Mustela vison	—	1	1	—	—	—	—	—	—	—	—	—	—	—	2
Taxidea taxus	—	1	1	10	—	—	1	—	—	—	—	—	—	—	13
Spilogale putorius	—	1	—	—	1	—	1	—	—	—	—	—	—	—	3
Mephitis mephitis	—	—	—	—	1	—	—	—	—	—	—	—	—	—	1
Lynx cf. *rufus*	—	—	—	1	1	—	—	—	—	—	—	—	—	—	2
Equus sp.	—	—	1	1	1	—	—	—	—	—	—	—	1	—	3
Camelops cf. *hesternus*	—	—	—	—	—	—	—	—	—	—	—	—	—	—	1
Odocoileus cf. *hemionus*	—	—	—	2	—	—	—	—	—	—	—	—	—	—	2
Antilocapra americana	—	—	—	5	—	—	—	—	—	—	—	—	6	—	11
TOTALS	212	286	189	1374	1069	0	428	1	158	0	8	0	145	7	3877

CHAPTER 2

The Basic Counting Units

There are certain basic kinds of questions that every faunal analyst asks of a set of bones and teeth from an archaeological site. Many of these questions are subsistence-related (what kinds of animals were utilized by the occupants of the site, and how did this utilization change through time?). Just as many are paleoenvironmental in orientation (what kinds of animals occurred in the area surrounding the site at the time the fauna accumulated, what kinds of changes occurred in the living fauna through time, and what inferences can be drawn concerning past environments in the area from this information?). Sitting in their labs, all faunal analysts go through the same basic procedures to answer these kinds of questions. They begin with a set of unidentified bones and teeth retrieved from various excavation units and, often, from various strata of a site. If they use the same procedures I use, they have each of these specimens numbered and catalogued, and then separate their material into specimens* that they can and cannot identify. Perhaps reserving the unidentifiable specimens for later work (though what is unidentifiable to one analyst is often not so to another, so that "unidentifiable" is an entirely different sort of category from, say, *Bos taurus*), they focus on the identifiable materials. They spend hours, days, or years identifying that material to the lowest taxon possible given their skill. At the end of this part of the process, they have lists of identified specimens with which they must now do something.

The first something that I do at this stage is to place these identified specimens into stratigraphic order. That is, I correlate each identified specimen with the stratum from which it came. Once this is done, I can answer some of the questions with which I began, since I now know what taxa comprise the faunal

---

* My use of the terms *specimen* and *element* follows Shotwell (1955, 1958): a *specimen* is a bone or tooth, or fragment thereof, from an archaeological or paleontological site, while an *element* is a single complete bone or tooth in the skeleton of an animal. Thus, a proximal humerus from an archaeological site is a specimen, while the humerus itself is an element.

assemblage* of each stratum. In particular, I can answer any questions I might have that require only that I know what taxa were present and what taxa were absent from any given stratum in the site.

Although presence/absence information of this sort can be put to good use (Grayson 1982), the questions faunal analysts ask routinely require much more. These questions demand that the absolute abundance of each taxon in each analytically important subdivision of the site be measured; often, these measures of absolute abundance (for instance, 100 bones or 100 individuals of Taxon A) are then transformed into measures of relative abundance (for instance, Taxon A forms 30% of the assemblage).

To get to the point where absolute and relative abundances can be measured may be time-consuming, difficult, frustrating, or boring. Usually, it is all of these. At least, however, reaching this point is a matter of rather straightforward technical expertise, of knowing which cusp on which tooth of which mouse bends which way. How to get beyond this point, how to measure abundance, is the subject of this chapter.

The Number of Identified Specimens

The basic counting unit that must be used in any attempt to quantify the abundances of taxa within a given faunal assemblage is the identified specimen, the single bone or tooth or fragment thereof assigned to some taxonomic unit. There are many measures that can be derived from a set of identified specimens. They can be transformed into minimum numbers of individuals, they can be weighed, they can be used to estimate the size of the death population (Fieller and Turner 1982), they can be transformed into animal weights (Reitz and Honerkamp 1983), and so on. But all of these measures start with the identified specimen.

For many years, the number of identified specimens (NISP) per taxon was used as the standard measure of taxonomic abundance within archaeological faunas. Bones from a given fauna were identified, numbers of identified specimens per taxon determined, and the NISP values themselves used to examine changing taxonomic frequencies through time and across space. Of the many examples of these kinds of studies, one stands out because of its pioneering

---

* I use the term faunal *assemblage* to refer to the entire set of faunal specimens from a given cultural or geological context, in which the defining context is provided by the analyst. Thus, all the faunal material from a single site can be referred to as a single assemblage, or that material can be divided into a series of assemblages depending on the analyst's goals. I use the term faunal *aggregate* and faunal *assemblage* interchangeably: as I discuss later in this chapter, the process of aggregation is the process of defining the boundaries for a given faunal assemblage.

nature and, not unrelated, because of the problems it displays. This is the classic study of the faunas of Tabūn and Wad caves by D. M. A. Bate (1937).

Tabūn and Wad caves are located on Mt. Carmel, some 3 km east of the modern shore of the Mediterranean Sea. Excavated by Dorothea Garrod during the late 1920s and early 1930s, these sites provided human bones and artifacts from late Pleistocene deposits that continue to play a major role in our interpretation of human biological and cultural history. The sites also provided large collections of vertebrate faunal remains.

These faunal remains were analyzed by Bate. A paleontologist, Bate knew precisely what she wished to do with these materials:

> An endeavour has been made to use this great collection of animal remains as a basis for a detailed history of the unfoldment of the faunal assemblages which succeeded each other during a not inconsiderable portion of the Pleistocene period in Palestine. This history of the fauna is inextricably interwoven with that of the changing climatic and environmental conditions, a fact which makes it possible to picture in broad outline some of the varying aspects of the country during this time. (Bate 1937:139)

Interest in faunal and climatic history led Bate not only into an examination of taxonomic frequencies much more detailed than was common at the time, but also led her to assume that any changes in frequencies that she discovered were, in fact, due to changes in the frequencies of those animals in the surrounding environment. The hindsight provided by 50 years of subsequent work shows that this assumption was probably inappropriate, but that does not detract from the historic importance of her research.

Bate was also admirably clear on how she conducted the details of her study:

> as a preliminary to the study of a species, individual specimens were marked in Indian ink with the name of the cave and of the Level from which they came. Exception was made only in the case of the smallest specimens, which were separately labelled in glass-topped boxes. This led to a table being made for each species recording the number of identified specimens found in the different Levels, even when these amounted to several thousands, as in the case of the Gazelles and *Dama mesopotamica*. (Bate 1937:139)

Bate then plotted changing percentages of identified specimens of her most abundant taxa, *Gazella* and *Dama*, across levels, and thus through time, at Tabūn and Wad (Figure 2.1). The results seemed impressive: there were major shifts in the relative abundances of these taxa. Deer, Bate noted, are animals of relatively moist climates; gazelles are animals of drier habitats. Accordingly, she interpreted times of high deer abundance as times of relatively moist climates, and times of high gazelle abundance as times of relatively dry climates. Although she did not plot changing relative abundances of other taxa through time in this fashion, she did note that the comings and goings of these other taxa were not inconsistent with her interpretation of the *Dama – Gazella* ratios.

THE NUMBER OF IDENTIFIED SPECIMENS

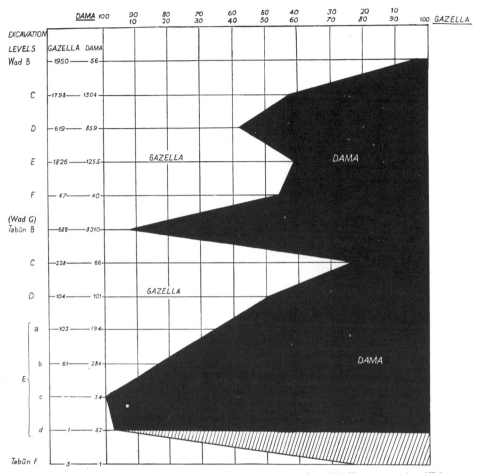

Graph showing the comparative frequency of *Dama* and *Gazella* during the period of human occupation of Tabūn and M. Wad. This is suggested as an indication of varying moist (*Dama*) and dry (*Gazella*) climatic conditions.
The actual numbers of specimens are given in the left-hand column.
The earliest part of the deposit is shown shaded owing to the very small number of specimens obtained.

Figure 2.1 Changing frequencies of *Gazella* and *Dama* at Tabūn and Wad, as calculated by Dorothea Bate from numbers of identified specimens. (From Bate 1937, courtesy of Oxford University Press.)

There are several problems with Bate's analysis, all of them instructive. The analysis suffers from the problem that plagues all closed arrays. Since percentages must sum to 100, the fact that the relative abundance of one taxon increases at the same time that the relative abundance of another decreases may be telling us more about the constraints imposed by closed arrays than about the behavior of the variables in which we are really interested. This difficulty was among the reasons that led pollen analysts to develop methods of studying absolute pollen influx rates (Davis 1967). Bate's study was particularly prone to such constraints because she was examining the relative abundances of only two taxa, and was thus dealing with a system in which there was only one degree of freedom: when the relative abundance of *Dama* increased, the relative abundance of *Gazella* had to decrease, and vice versa. In addition, Bate's argument suffers from her underlying assumption that the remains of animals accumulated in her sites in at least rough proportion to their abundance in the surrounding environment. As a result, Bate did not give serious consideration to the possibility that causes other than climatic change could have caused the variations in relative abundances that she had detected. Bate's assumption was the paleontological equivalent of the still-common archaeological assumption that if bones are present in a deposit that also contain archaeological debris, those bones must relate to human hunting patterns. Just as archaeological assumptions often seem weak to paleontologists, paleontological assumptions often seem weak to archaeologists, and Vaufrey (1939) was quick to point out that changing human hunting patterns might have played a role in causing the variations in relative abundance that Bate had found, a view that has found increasing support in recent times (Jelinek 1982). In short, problems relating both to the methods of quantification and to taphonomy clouded the meaning of Bate's results.

Bate's important study shows the simplicity of analyses of taxonomic frequencies based on specimen counts. The analyst identifies bones and teeth, counts the number of identified specimens per taxon, and analyzes the resulting numbers. By the early 1950s, however, it was becoming clear to some faunal analysts that specimen counts were liable to bias on many counts. Indeed, after the popularization of the use of minimum numbers of individuals in archaeological work during the mid-1950s, the perceived number of flaws that lurked within specimen counts grew rapidly. By the 1970s, any faunal analysis that did not at least present minimum numbers was suspect, although one that presented only minimum numbers was not.

Of the many criticisms that have been directed toward specimen counts, 11 have been cited most frequently:

1. Numbers of identified specimens are affected by butchering patterns, so that differences in specimen counts per taxon may simply reflect the fact that some animals were retrieved from kill sites whole, while others were butchered on the spot with only selected portions retrieved (for an important discussion, see Binford 1978, 1981). As White (1953:396–397) noted, "with large animals like the bison, most of the metapodials, pelves, vertebrae and skulls are left at the kill, while deer and antelope were probably brought back to the village for butchering." Perkins and Daly (1968) later named this postulated phenomenon the *schlepp effect*,

THE NUMBER OF IDENTIFIED SPECIMENS

from the German verb *schleppen*, to drag. The effect is extremely well documented ethnographically (Binford 1978, 1981; Read-Martin and Read 1975).

2. Numbers of identifiable specimens vary from species to species, so that differences in numbers of specimens per taxon may simply reflect the fact that, for instance, an analyst can identify all the specimens of *Bison* in a fauna, but can only identify the cranial elements and teeth of a small rodent to the same taxonomic level. Thus, Shotwell (1958) noted that he was able to identify 133 elements of the skeleton of late Tertiary rhinos, while he could identify only 43 elements of the skeletons of late Tertiary rabbits. In a situation of this sort, a fossil fauna that contained 1290 identified specimens of the rhinoceros *Aphelops* and 430 identified specimens of the rabbit *Hypolagus* would contain three times as many identified specimens of rhino than of rabbit, but those numbers could not be used in a straightforward way as an estimator of the relative abundances of the animals in either the fossil fauna or in the prehistoric living fauna.

3. The use of NISP assumes that all specimens are equally affected by chance or by deliberate breakage. A butchering technique that partitioned the bones of large mammals into more pieces than the bones of small ones, for instance, would provide specimen-per-taxon counts that reflected butchering techniques more than they reflected the numbers of the animals that contributed to the fauna (Chaplin 1971). This phenomenon is well described in both the ethnoarchaeological (Binford 1978) and the archaeological literature (Guilday *et al.* 1962).

4. Differential preservation affects the number of identifiable specimens per taxon, so that the numbers identified by an analyst today may bear an unknown relationship to the numbers originally deposited. Guthrie (1967), for example, noted that within four late Pleistocene Alaskan faunas, second phalanges of bison were more abundant than the larger third phalanges, and suggested that poorer preservation of the "more porous and fragile" (1967:243) second phalanges accounted for the difference. Brain (1969) demonstrated that the preservation of skeletal parts in a collection of modern bones from Hottentot villages was positively correlated with the specific gravity of each skeletal part, and inversely correlated with the length of time required for the epiphyseal union of each part. In the most detailed study of this sort done to date, Lyman (1982b) used a photo densitometer to measure the density of modern skeletal elements, demonstrating that bone survivability is strongly correlated with bone density (see also Binford and Bertram 1977). There can be no doubt that preservation affects not only the number of specimens that remain to be identified for a given taxon

within a given fauna, but that preservation differentially affects the bones of different taxa as well.
5. The number of identified specimens can give misleading results when one or more taxa are represented by entire individuals, while other taxa are represented only by disarticulated and fragmented bones and teeth (Bökönyi 1970; Chaplin 1971; Payne 1972a). In a collection in which a given taxon is represented by 200 complete right femora while a second taxon is represented by 200 articulated elements of a single skeleton, counts of identified specimens may give a very biased picture of relative taxonomic abundance.
6. For many of the reasons noted above, a number of authors have concluded that the use of the number of identified specimens leads to difficulties "in statistical treatment caused by sample inflation" (Payne 1972a:68; see also Chaplin 1971; Watson 1979). An individual animal once represented by 50 identifiable bones at a site may, through time, come to be represented by 200 specimens as various processes fracture the bones originally deposited. A single bison skull may come to be represented by 30 isolated teeth and skull fragments. Even if this process is applied equally to all taxa in the fauna, the increase in sample sizes that are involved can lead to significant differences in the results of statistical tests applied to these data (Payne 1972a). Table 2.1 demonstrates the effect on a simple χ^2 test. In Table 2.1A, 30 specimens of *Bos* and 40 of *Ovis* are originally deposited in Stratum 1, and 40 of *Bos* and 30 of *Ovis* in Stratum 2, of an archaeological site. The differences in taxonomic representation between these strata are not statistically significant ($\chi^2 = 2.86$, $p > .05$). But, should some process — for instance, trampling by person or beast — fragment each set of remains equally, very different results can be obtained. In Table 2.1B, each original bone has been broken 10 times; now, the differences between strata are highly significant ($\chi^2 = 28.57$, $p < .01$).
7. The unit may be affected by collection techniques. Samples collected without screening will contain primarily large specimens compared to those collected with screening; samples collected with fine mesh screen will contain a higher proportion of smaller specimens than samples collected with larger mesh screen. This phenomenon is well studied, and differentially affects numbers of specimens retrieved both within and among taxa (Brain 1967; Casteel 1972, 1976a; Payne 1972b; Thomas 1969; Watson 1972; also see Chapter 6 below).
8. The number of identified specimens cannot, by itself, address questions of biomass, and meat weights are often of far greater importance in examining prehistoric economies than is the number of bones by which a given taxon is represented: "it is really meat we are interested in, not

THE NUMBER OF IDENTIFIED SPECIMENS

TABLE 2.1
An Example of the Effects of Bone Fragmentation on the Statistical Assessment of Significant Differences between Faunas

A. NUMBER OF IDENTIFIED SPECIMENS PER TAXON, INITIAL FAUNA[a]		
	Stratum 1	Stratum 2
Bos	30	40
Ovis	40	30

B. NUMBER OF IDENTIFIED SPECIMENS PER TAXON, FRAGMENTED FAUNA[b]		
	Stratum 1	Stratum 2
Bos	300	400
Ovis	400	300

[a] $\chi^2 = 2.86$, $p > .05$
[b] $\chi^2 = 28.57$, $p < .01$

bones" (Daly 1969:148). This particular criticism observes that meat weights cannot be obtained directly from bone counts in most settings, and that taxonomic frequencies based on bone counts may bear little relationship to the dietary contribution made by those taxa. Equal numbers of bison and rabbit bones do not imply that those animals were of equal dietary importance (Bökönyi 1970; Daly 1969; Uerpmann 1973a).

9. Because of the cumulative weight of the problems discussed to this point, Chaplin (1971) has argued, numbers of identified specimens do not allow valid comparisons between faunas (and presumably between any analytic units), and should be abandoned.

10. Even if NISP values did allow valid comparisons, Chaplin (1971) has also argued, specimen counts simply do not support as many analytic techniques as the minimum number of individuals, and should be abandoned for this reason as well.

11. Finally, numbers of identified specimens have been criticized because of the potential interdependence of the units being counted. Given current knowledge, there is no way of demonstrating which bones and teeth and fragments of bones and teeth necessarily came from different individuals across an entire faunal assemblage, and thus no way of resolving the patterns of specimen interdependence that must surely characterize many specimen samples. Since the statistical methods used to analyze these samples — from simple counts to percentages to chi-square and

beyond—assume not only that the items being manipulated are representative of the population about which inferences are being made, but also that each item counted is independent of every other item counted, the application of statistical methods to NISP-based faunal counts is inappropriate (Grayson 1973, 1977b, 1979b).

These criticisms of specimen counts are a mixed and overlapping lot. It is interesting to note that these criticisms largely arose after minimum numbers had already become well-accepted as a counting unit, and thus were used to justify minimum numbers, rather than to develop a new method of counting. Nonetheless, this set of criticisms has been so effective that it is now rare to find an analysis based entirely on counts of identified specimens. Today, the vast majority of faunal studies employ minimum numbers of individuals as the basic unit of quantification, while presenting but not analyzing specimen counts. At one time, my own assessment of specimen counts was highly negative, and my early faunal work analyzed either minimum numbers alone (e.g., Grayson 1976) or both minimum numbers and specimen counts simultaneously (e.g., Grayson 1977b), but never considered specimen counts alone (for just such an approach, see Grayson 1983b).

How compelling are these criticisms of numbers of identified specimens? I will argue that they are not sufficiently compelling to justify the dismissal of NISP values as a counting unit, but I must discuss minimum numbers in some detail before making that argument. It is worth emphasizing at this point, however, that because criticisms of specimen counts developed largely after minimum numbers had become widely accepted, and were thus primarily used to justify minimum numbers rather than to build a new set of methods from the ground up, a simple list of the most common objections strongly implies that specimen counts are more troublesome than they really are. Many of the published criticisms really address only one issue, interdependence (for instance, Numbers 3, 5, and 6 above). Others present problems that are readily remedied (for instance, Numbers 2 and 7), while still others do not address substantial issues (for instance, Numbers 8 and 10).

Criticisms of specimen counts that assert that those counts are of little use because they do not allow the assessment of meat weights, or in general do not support a wide variety of derived analytic procedures, do not address the issue of whether specimen counts are a valid measure of the taxonomic composition of a given fauna. Certainly, no one would argue that measures of taxonomic frequency based on specimen counts can be used as measures of all other variables in which a faunal analyst might be interested; that they do not allow the measure of such variables does not imply that they are not valid measures of taxonomic composition per se. It is incorrect to dismiss specimen counts as a measure of taxonomic abundance simply because all measurement goals can-

THE NUMBER OF IDENTIFIED SPECIMENS   25

not be achieved with them, and it is certainly inappropriate to employ minimum numbers of individuals simply because more can be done with them, without assessing the possibility that minimum numbers themselves might be very seriously flawed.

Other criticisms of specimen counts can be removed through the application of statistical or field methods to ensure valid comparability. The criticism afforded by collection procedures is a case in point. Rather than dismissing a measurement unit because our means of collecting data are flawed, we should focus instead on developing retrieval procedures that either eliminate the problem, or at least allow us to measure the magnitude of that problem (e.g., Thomas 1969). That different species have different numbers of identifiable elements also poses no insurmountable problem. Shotwell (1955, 1958, 1963) remedied this difficulty in straightforward fashion by norming specimen counts with numbers of identifiable elements per taxon. In addition, since faunal analysts are generally interested in studying changes in *relative* taxonomic abundances across space or through time, and since the number of identifiable elements per taxon is, or should be, a function of the particular taxa being identified, this critique applies only to single faunal assemblages. Relative abundances among assemblages will not be adversely affected as long as the same set of taxa is being examined (thus, Bate's analysis is immune from this problem). It is also true that any critique of specimen counts on this score applies as well to minimum numbers. That it has been applied only to minimum numbers comes from the fact that this criticism, like most of the others, has come not from an attempt to build a sound set of faunal methods, but instead from an attempt to argue for the use of minimum numbers.

Much the same case could be made for objections that stem from the differential preservation of elements. There can be no doubt that bones are differentially preserved, that bone density plays a crucial role in mediating bone preservation, and that the effects of bone density are sufficiently pronounced that even proximal and distal ends of the same long bones will be differentially preserved. The work of Brain (1969, 1976, 1981) and Lyman (1982b), among others, establishes that point well. Differential preservation will affect the bones of all taxa in all faunas, and this will affect all specimen counts. However, differential preservation will affect the most abundant elements that are the basis of minimum number determination (see below) as well as they affect all other elements. Again, this objection is leveled against specimen counts primarily to justify the use of minimum numbers, not to construct a new set of methods.

Many of the remaining objections have a common, and valid, theme. Concerns with differential breakage, the effects of articulated sets of bones, sample size inflation, and butchering-related differential breakage all have at their heart concern with the effects of the lack of independence among the units being counted. As I have noted, this problem is a serious one, since all counting procedures that we employ assume that the items being counted are not mechanically interdependent. How, then, can we assure ourselves that we are not counting the same thing more than once? How can we be sure, for instance, that when we tally 20 specimens for Taxon A and 100 specimens for Taxon B that both sets of material did not result from one individual animal of each taxon? The answer, unfortunately, is that we cannot know. It can, of course, be assumed that each specimen was necessarily contributed by a different individual, as was explicitly done by Hesse and Perkins (1974) and Gilbert and Steinfeld (1977), and implicitly by most others who use specimen counts to quantify taxonomic abundances. Clearly, however, this assumption does not address the underlying issue. Assuming independence does not create independence among the units being counted. If this assumption seems unreasonable, it might instead be assumed that interdependence is randomly scattered across all taxa (Grayson 1979b; see also Damuth 1982). In either case, the difficulty becomes demonstrating that the assumption is an appropriate one. Minimum numbers, as I shall discuss, can solve this problem, but do it by creating even more intractable difficulties.

Real challenges to the use of specimen counts as a measure of relative taxonomic abundance are also posed by the differential deposition of varying parts of animals in a site. Such differential deposition can be created by the schlepp effect, and by a wide range of other mechanisms, including the activities of scavengers and carnivores. While detailed taphonomic studies (e.g., Andrews and Evans 1983; Binford 1981) can help control for the effects of such processes, we are no closer to full control here than we are to controlling for the effects of bone density on bone preservation. Butchering and similar processes do provide valid objections to the use of specimen counts as a measure of taxonomic composition.

Thus, many of the objections that have been leveled at specimen counts as a measure of the taxonomic composition of vertebrate faunas from archaeological sites are not sufficiently powerful to suggest that these counts should be abandoned. Among these, I include objections stemming from differential element identifiability and differential element preservation. Others, such as the inability of specimen counts to support a wide range of analytic techniques, do not address the validity of specimen counts as a measure of taxonomic abundance per se. However, the problems posed by specimen interdependence and by the alteration of the numbers of specimens deposited on sites as a result of such extrinsic processes as butchering and the activities of carnivores do provide substantial objections. Before assessing precisely how substantial these objections are, the other common quantifier of taxonomic abundance must be addressed.

The Minimum Number of Individuals

The potential effects of interdependence on specimen counts suggests that an alternative unit not affected by this problem be sought for the quantification of taxonomic abundances within vertebrate faunas. The minimum number of individuals (MNI) per taxon can be calculated in such a way as to possess this quality.

Even though minimum numbers of individuals solve the potentially severe problems of specimen interdependence, the introduction of minimum numbers into the archaeological literature by White (1953) was made for a very different reason. White (1953) was struck by the fact that archaeological sites in the North American Great Plains yielded faunas that differed dramatically from one another. "Some groups," White (1953:396) noted, "set an extremely 'varied' table, while others appear to have subsisted almost entirely on one species of food animal." This being the case, White became interested in measuring these differences, in determining "the percentage which each species contributes to the diet of the people" (1953:396).

White rejected the use of specimen counts for two reasons. First, he noted that butchering techniques would probably result in the differential deposition of body parts on sites, as I have mentioned. Second, he recognized that differences in the sizes of hunted species meant that each species did not contribute equally to the diets of the people involved, and that specimen counts did not directly address this issue: "four deer," White (1953:397) observed, "will be required to provide as much meat as one bison cow," and simple specimen counts did not take this fact into account. Of these two objections to numbers of identified specimens, White was clearly most impressed by the second, since his goal was "to determine the amount of meat furnished by any given species" (1953:397).

In place of specimen counts, White recommended the use of "the number of individuals" per taxon represented in the faunal sample, now routinely called the minimum number of individuals. White's approach to calculating these numbers was plainly stated:

> The method I have used in the studies on butchering techniques is to separate the most abundant element of the species found . . . into right and left components and use the greater number as the unit of calculation. This may introduce a slight error on the conservative side because, without the expenditure of a great deal of effort with small return, we cannot be sure all of the lefts match all of the rights. (1953:397)

It was no accident that White, who presented the first well-reasoned introduction of minimum numbers into archaeology of which I am aware, was a paleontologist. Vertebrate paleontologists had long used minimum numbers in an identical fashion. The censuses of the mammals and birds of Rancho La Brea published during the 1920s and 1930s, for instance, had been conducted with minimum numbers (e.g., Howard 1930; Merriam and Stock 1932; Stock 1929), and discussions of the numbers of individuals represented by a set of fossils are common in the nineteenth-century literature (e.g., Buckland 1823). After White defined minimum numbers in 1953, others introduced variations into how those numbers were to be defined. Flannery (1967), for instance, spent the "great deal of time with small return" that White had noted was needed to determine if all the rights matched all the lefts, adding extra individuals if they did not match. Although techniques for defining minimum numbers have become more refined over the years (e.g., Nichol and Creak 1979), the basic approach has remained much the same since White's work appeared.

White's discussion of the minimum number of individuals makes the determination of those numbers appear simple. An archaeological site is excavated and the bones and teeth from that site retrieved and identified. Then, the identified materials are distributed into such analytic units as strata and house floor debris, creating smaller clusters or aggregates (I use the terms interchangeably) of faunal specimens. Finally, the operational definition of minimum numbers is applied separately to each taxon represented in each of these aggregates of faunal material. The resulting numbers are then manipulated in any further analysis.

Minimum numbers determined in this way would seem to have a number of advantages over specimen counts. Most obviously, the numbers determined for any given faunal aggregate are all independent of one another. Twenty left femora of one species in a given cluster of faunal material must all have come from different individuals, and there can be no doubt that the same thing is not being counted more than once within that cluster. This is a major gain, since it eliminates the most severe disadvantage of specimen counts, interdependence, as regards each separate faunal cluster. In addition, and as White (1953) appropriately observed, minimum numbers can diminish the effect of differential retrieval of bone material from a kill site. If only the long bones of bison were brought back to an occupation site while entire skeletons of deer, antelope, and rabbits were retrieved, minimum numbers would not be affected, but specimen counts would be.

If this were all there were to it, minimum numbers would seem to possess many of the attributes required from a measure of taxonomic abundance. Although certain problems would remain — for instance, differential preservation might still cause problems — at least minimum numbers would solve the problems related to interdependence that confront specimen counts. Certainly, the fact that minimum numbers have been so widely adopted by archaeologists (and by many paleontologists) during the past three decades would suggest that they do, in fact, provide a relatively trouble-free measure of taxonomic abundance.

THE MINIMUM NUMBER OF INDIVIDUALS 29

Unfortunately, this is not the case. The simple operational definition of minimum numbers glosses over the crucial stage in defining those numbers: the definition of the clusters of faunal material from which minimum numbers are defined. It is easy to demonstrate that the numerical values of minimum numbers of individuals vary with the way in which faunal material from a given site is divided into those smaller aggregates. Not only may the use of different approaches to aggregation change the calculated minimum numbers, but these changes in abundance will probably occur differentially across taxa. Indeed, a highly motivated investigator can at times apply different approaches to aggregation to the same set of faunal materials in such a way as to obtain a wide range of outcomes, and then select the set of aggregates that provides the most impressive support for any given hypothesis. Given that it is usually impossible to recalculate minimum numbers using different approaches to aggregation without the raw data, and often without the site records, at hand, there is generally no way for the reader of any given faunal report to know how minimum numbers would have differed had different approaches to aggregation been employed. There are no such difficulties with specimen counts. I noted this phenomenon a decade ago (Grayson 1973); Casteel (1976/1977) later observed that this same effect had been discussed by the Russian faunal analyst Paaver in 1958. Given the possible magnitude of aggregation effects and the implications of those effects for the use of minimum numbers, I will explore them more fully here.

The fact that different aggregation techniques applied to a single faunal collection may produce minimum numbers that are widely different can be readily understood through examination of the process of minimum number calculation. Recall that, as White (1953) noted, the basic step in minimum number determination is the specification of the "most abundant element" for any given taxon. This specification can be accomplished only after some decision has been made as to precisely how a given faunal collection is to be subdivided into separate analytic units. If all the faunal material from the site is to be treated as a single large aggregate, the most abundant element will be defined once per taxon for that collection. As the collection is divided into smaller and smaller aggregates of faunal material — for instance, by subdividing the collection according to the strata or vertical excavation units from which it came — the number of separate specifications of most abundant elements will increase. Thus, dividing a faunal collection into a smaller number of larger faunal aggregates will lead to the definition of smaller absolute minimum number values then will dividing the same collection into a larger number of smaller aggregates. The extremes of the process are easily defined. The smallest possible minimum number values will result when the entire faunal collection is treated as a single large faunal aggregate. Here, most abundant elements are defined only once for each taxon. The largest possible minimum number values will result when the spatial boundaries of each aggregate are so small as to contain only a single specimen. Here, each specimen becomes a "most abundant element," because it is the only specimen in the faunal aggregate, and the minimum number of individuals will equal the number of identified specimens per taxon, the highest value it can attain.

Table 2.2 presents a contrived example of the effects of aggregation on minimum number values. Table 2.2A shows the distribution of 350 identified specimens across two taxa in a fauna treated as a single aggregate. Taxon 1 is represented by 110 specimens, while the most abundant element, the right femur, defines an MNI of 50 for this taxon. Taxon 2 is represented by 240 specimens, while the most abundant element, again the right femur, defines an MNI of 100 for this taxon. In Table 2.2B, I have presented the same fauna distributed across two strata. Taxon 1 is still represented by 110 specimens, but there are now two most abundant elements, the right humerus in Stratum 1 and the right femur in Stratum 2. The total minimum number of individuals has risen to 65. Taxon 2 is still represented by 240 specimens, but there are now two most abundant elements here as well, the right femur in Stratum 1 and the left femur in Stratum 2. The total minimum number of individuals has risen to 130. Dividing the one, large faunal aggregate into two smaller aggregates has added 45 individuals to our collection; the creation of smaller and smaller aggregates may continue to add individuals until NISP = MNI for each taxon.

In the example provided in Table 2.2, different aggregation methods have added minimum numbers to both taxa, but have affected those taxa identically. In the initial aggregation, Taxon 2 (with an MNI of 100) was twice as abundant as Taxon 1 (with an MNI of 50). This ratio remained the same when the fauna was redistributed according to the two strata from which it came (an MNI of 130 for Taxon 2, compared with an MNI of 65 for Taxon 1). Were this the only effect of aggregation, faunal analysts might have little to worry about, since minimum numbers are rarely treated as absolute values meaningful in-and-of themselves, but instead take on meaning only in relationship to other minimum number values. Unfortunately, the example in Table 2.2 was contrived, and the changes in minimum numbers across taxa that occur when different approaches to aggregation are employed almost always differentially alter the calculated abundances of taxa. If, for instance, Taxon A is twice as abundant as Taxon B under one approach to aggregation, it is not likely that it will be twice as abundant under a different approach. This is true because different aggregation methods specify different most abundant elements, and because the most abundant elements for one taxon will in almost all instances be spatially distributed differently from the most abundant elements of all other taxa. Only if all elements that enter into the calculation of minimum numbers are distributed in identical ways across all aggregation units will different approaches to aggregation fail to differentially alter calculated minimum number abundances among

THE MINIMUM NUMBER OF INDIVIDUALS

TABLE 2.2

The Effects of Aggregation on Minimum Numbers: Abundance Ratios Unaltered

A. COLLECTION TREATED AS A SINGLE AGGREGATE

Taxon 1	Taxon 2
50 right femur	100 right femur
40 right humerus	80 right humerus
20 left humerus	60 left femur

Taxon 1: NISP = 110
MNI = 50
Taxon 2: NISP = 240
MNI = 100

B. COLLECTION TREATED AS TWO AGGREGATES DIVIDED ACCORDING TO STRATIGRAPHIC PLACEMENT

	Taxon 1	Taxon 2	
Stratum 1	25 right femur 40 right humerus	100 right femur 80 right humerus 30 left femur	MNI, Taxon 1 = 40 MNI, Taxon 2 = 100
Stratum 2	25 right femur 20 left humerus	30 left femur	MNI, Taxon 1 = 25 MNI, Taxon 2 = 30
Σ NISP	110	240	Σ MNI, Taxon 1 = 65 Σ MNI, Taxon 2 = 130

taxa. The example in Table 2.3 makes this effect clear. Unlike the example presented in Table 2.2, the effects shown in Table 2.3 are not contrived.

Table 2.3A presents the number of identified specimens for two taxa in a small fauna treated as a single aggregate. Taxon 1 is represented by 55 specimens. The single most abundant element for this taxon, the right humerus, defines an MNI of 30. Taxon 2 is represented by 70 specimens. The single most abundant element for this taxon, again the right humerus, defines an MNI of 40. In Table 2.3B this small fauna has been aggregated according to the three strata from which it came. Now, the most abundant element is defined three separate times for each taxon, and the total minimum number values are 47 for Taxon 1 and 40 for Taxon 2. The absolute minimum number counts have increased because, by subdividing the single faunal aggregate into several smaller ones, the number of specifications of most abundant elements has increased. The ratios of the abundances of these taxa to one another have changed dramatically (from $\frac{30}{40}$, or 0.75, in Table 2.3A to $\frac{47}{40}$, or 1.18, in Table 2.3B) because the spatial distribution of the elements that play a role in minimum number determination differs dramatically between the taxa. Although I have used natural strata to subdivide the faunal collection presented in Table 2.3, it should be clear that aggregation effects may be present no matter what kind of analytical unit

TABLE 2.3

The Effects of Aggregation on Minimum Numbers: Abundance Ratios Altered

A. COLLECTION TREATED AS A SINGLE AGGREGATE

Taxon 1	Taxon 2
25 left humerus	30 left humerus
30 right humerus	40 right humerus
Taxon 1: NISP = 55	
MNI = 30	
Taxon 2: NISP = 70	
MNI = 40	

B. COLLECTION TREATED AS THREE AGGREGATES DIVIDED ACCORDING TO STRATIGRAPHIC PLACEMENT

	Taxon 1	Taxon 2	
Stratum 1	22 left humerus	20 left humerus	MNI, Taxon 1 = 22
	5 right humerus	20 right humerus	MNI, Taxon 2 = 20
Stratum 2	3 left humerus	0 left humerus	MNI, Taxon 1 = 17
	17 right humerus	10 right humerus	MNI, Taxon 2 = 10
Stratum 3	0 left humerus	10 left humerus	MNI, Taxon 1 = 8
	8 right humerus	10 right humerus	MNI, Taxon 2 = 10
Σ NISP	55	70	Σ MNI, Taxon 1 = 47
			Σ MNI, Taxon 2 = 40

(house pits, storage pits, arbitrary levels, and so on) is involved. Absolute minimum number values increase as the sample is more finely divided, until MNI = NISP for each taxon, and increase differentially as long as the most abundant elements among taxa are not spatially distributed in identical ways.

A general statement about the effects of aggregation may be made if the distribution of element frequencies is conceived of as a series of peaks and valleys across possible aggregation units. For any aggregation unit, it is the peak of the distribution of the most abundant element that defines the minimum number of individuals. If the distributions of elements across taxa are identical in form, then different approaches to aggregation will provide different absolute minimum number values, but the ratios of one taxon to another will remain constant. If, on the other hand, the distributions of most abundant elements are not identical in form and peaks of most abundant elements occur differentially across aggregation units, different approaches to aggregation will not only define different absolute taxonomic abundances, but will also alter the relative abundances of those taxa (see Figure 2.2).

Thus, two general situations may be specified: that in which the distributions of most abundant elements of all taxa are identical, and that in which the distributions of most abundant elements differ among taxa and the frequency

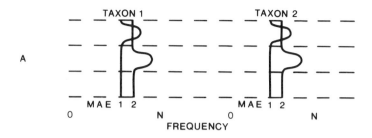

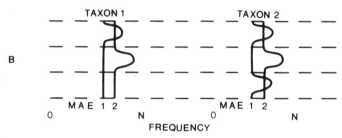

Figure 2.2 Distributions of most abundant elements (MAE) and the effects of aggregation. Dashed lines represent borders of possible aggregation units. A, Distributions of most abundant elements identical; different aggregation methods produce different absolute MNI values but identical ratios of abundance. B, Distributions of most abundant elements differ between taxa, with frequency peaks of most abundant elements differentially spaced across aggregation units; different aggregation methods produce different absolute MNI values and different ratios of abundance.

peaks of those elements are differentially spaced across aggregation units. In the former case, the absolute values of minimum numbers will change when different aggregation approaches are employed, but relative abundances will not. In the latter, and usual, case, both absolute and relative abundances based on minimum numbers will change as approaches to aggregation change. In all cases, of course, the number of identified specimens per taxon remains the same.

The nature of the distribution of most abundant elements is an empirical question and must be determined in each case. In fact, much might be learned from analysis of these distributions, above and beyond that gained by analyzing their effects on minimum numbers. Unfortunately, since precise horizontal and vertical locational data are rarely recorded for all faunal remains within a site, continuous distributions of specimens within a site can rarely be determined. However, even with general provenience data, distributions of elements across possible aggregation units can be discerned by using the number of each identified element per taxon to provide the peaks for each aggregation unit. If the distributions of most abundant elements for all taxa are identical, then minimum numbers may be used without concern for the effects of aggregation on ratios of taxonomic abundance. If the distributions are not identical, they may not be so used. There is, of course, an easier way to determine whether or not minimum numbers should be used to provide ratios of taxonomic abundance: calculate minimum numbers using different aggregation units, and then use the resultant numbers to determine ratios of taxonomic abundance. If the ratios change across aggregation methods, they are reflecting both taxonomic abundances within the site and the differential distribution of most abundant elements.

How often will the distribution of most abundant elements be identical across all taxa and across all aggregation units? Except in the trivial instance in which the collection consists of a single taxon, it is difficult to see that this will occur commonly. What is known about the taphonomy of archaeological faunas suggests that there is no reason to expect that element distributions across aggregation units will be identical. The effects of element preservation, butchering techniques, archaeological recovery techniques, the complexities of human behavior, the wide range of nonhuman mechanisms that introduce faunal material into sites, and so on, ensure that these distributions are likely to differ among taxa. After all, patterning of this sort is one of the variables in which faunal analysts are interested.

The implications of this situation should be clear. Absolute abundances indicated by minimum numbers are dependent on aggregation method; ratios of taxonomic abundance based on those numbers are dependent on the nature of the distribution of most abundant elements within the site. As a result, statistical analyses that require a ratio measurement scale can rarely be assumed to apply to minimum number data.

The Effects of Aggregation on Minimum Number Abundances: Two Examples

To illustrate some of the effects of aggregation on minimum numbers of individuals, I will examine two faunas using minimum number analysis. One of these examples deals with a fairly simple fauna — few taxa and few aggregation units — and has been published elsewhere (Grayson 1979b). I present it first because of its simplicity. The second example is drawn from Hidden Cave, and deals with a greater number of taxa represented by a larger sample of identified specimens that are distributed over a larger number of aggregation units.

Connley Cave No. 4 Connley Cave No. 4 (35 LK 50/4) is one of a series of six contiguous rockshelters located in the Northern Great Basin of south-central Oregon. Excavated by the late Stephen Bedwell in 1967, this site provided 1081 mammalian specimens that I was able to identify (Grayson 1977A, 1979A; see Table 2.4). The site was excavated by 10-cm arbitrary levels within each of four

TABLE 2.4

Numbers of Identified Specimens by Stratum, Connley Cave No. 4 Mammals[a]

	Stratum				
Taxon	1	2	3	4	Total
Lepus spp. (1)	7	3	256	388	654
Sylvilagus nuttallii (2)	12	5	75	27	119
Sylvilagus idahoensis (3)	—	—	43	34	77
Neotoma cf. cinerea (4)	3	3	40	9	55
Thomomys talpoides (5)	4	—	31	8	43
Bison bison (6)	—	—	8	19	27
Sylvilagus sp. (7)	1	1	8	12	22
Cervus elaphus (8)	—	—	8	13	21
Ochotona princeps (9)	—	—	7	6	13
Odocoileus cf. hemionus (10)	—	—	10	2	12
Dipodomys ordii (11)	—	1	4	2	7
Marmota flaviventris (12)	—	—	3	2	5
Canis cf. latrans (13)	—	—	1	4	5
Lynx cf. rufus (14)	—	—	1	3	4
Vulpes vulpes (15)	—	—	2	2	4
Microtinae (16)	—	—	1	3	4
Erethizon dorsatum (17)	—	—	1	2	3
Mustela frenata (18)	—	—	2	—	2
Gulo luscus (19)	—	—	—	1	1
Antilocapra americana (20)	—	—	1	—	1
Ovis canadensis (21)	—	—	1	—	1
Spermophilus beldingi (22)	—	—	—	1	1
					1081

[a] Numbers in parentheses are used to identify taxa in Tables 2.5–2.8; taxa are presented in order of specimen counts.

natural strata; faunal materials were recovered with the use of ¼-inch (.64-cm) screens. The entire sequence spans the period of 11,200 to 3000 B.P., with a conspicuous gap in radiocarbon dates between 7200 and 4400 B.P (Bedwell 1969, 1973).

I calculated minimum numbers for this fauna following an operational definition similar to that used by White (1953), in which all the specimens for a given taxon were separated into left and right components, and the larger of the two values used as the minimum number of individuals (in no case did an unpaired element reach sufficient numbers to act as a most abundant element). In addition, I followed Flannery (1967) in expending the time and effort to use patterns of epiphyseal union and other age indicators in defining minimum numbers.

In calculating minimum numbers for the Connley Cave No. 4 fauna, I used three different approaches to aggregation:

1. All faunal materials were separated by natural stratum and were then subdivided by 10-cm units within each stratum. The operational definition of minimum numbers was then applied to each of the 34 faunal aggregates so defined; the resultant values are symbolized as MNI_{10cm}.
2. All faunal materials were grouped according to the natural stratum from which they had come, and the operational definition of minimum numbers was then applied to each of the resulting four clusters of faunal material ($MNI_{stratum}$).
3. The entire sample of faunal material was treated as one large aggregate, and the operational definition of minimum numbers applied to this single, large faunal aggregate (MNI_{site}).

The resultant minimum numbers are shown in Table 2.5.

Table 2.5 illustrates the expected effects of changing approaches to aggregation on minimum number values. Dividing the faunal sample into the largest number of aggregates (34), each containing a relatively small fraction of the sample, defines the largest minimum number of individuals (254) for the entire collection. Treating the faunal sample as a single aggregate defines the smallest minimum number of individuals (109) for the whole collection. This situation results from the fact that the most agglomerative approach to aggregation — MNI_{site} — specifies only one element per taxon as most abundant, while more and more most abundant elements are defined as the approach to aggregation becomes increasingly divisive.

The differences between the absolute values of minimum numbers of individuals determined by the most divisive (MNI_{10cm}) and the most agglomerative (MNI_{site}) methods applied to the Connley Cave No. 4 mammals vary from minor to pronounced, depending upon the taxon involved. The maximum values that these differences can reach within any given faunal collection are easily determined. The smallest possible minimum number values are obtained when the entire assemblage is treated as a single aggregate, from which minimum numbers are determined. The largest possible minimum number values are obtained when the assemblage is divided into the largest possible number of aggregates, functionally attained when each specimen contributes an individual. Thus, as I have already noted, the largest minimum number values possible for any given assemblage are equal to the number of identified specimens for that assemblage. Since this is the case, maximum possible differences for minimum number values may be calculated by subtracting MNI_{site} from NISP for each taxon. The values that result for the Connley Cave No. 4 mammals when this is done are provided in Table 2.6. Because maximum possible differences in minimum number values necessarily decrease as the number of identified

THE MINIMUM NUMBER OF INDIVIDUALS

TABLE 2.5
Total Minimum Numbers of Individuals by Aggregation Method, Connley Cave No. 4 Mammals

Taxon	MNI_{10cm}	$MNI_{stratum}$	MNI_{site}
1	72	39	36
2	38	15	10
3	23	8	9
4	23	13	8
5	23	14	13
6	9	2	2
7	12	8	5
8	9	3	2
9	9	7	4
10	5	2	1
11	7	4	4
12	3	3	2
13	5	2	1
14	2	2	1
15	3	2	1
16	3	3	3
17	2	2	1
18	2	2	2
19	1	1	1
20	1	1	1
21	1	1	1
22	1	1	1
	254	135	109

specimens decreases, the effects of different methods of aggregating faunal data on absolute values of minimum numbers will be most pronounced for larger samples, least pronounced for smaller ones. That is, as specimen samples increase in size, the range in values that minimum numbers may take as a result of different aggregation methods increases, and the faith that may be placed in the meaning of absolute minimum number values decreases.

It is important that the meaning of the values presented in Table 2.6 be clear. If an analyst wanted to express the abundance of hares, *Lepus* spp., at this site in terms of minimum numbers of individuals, the values that this figure might take range from a minimum of 36 to a maximum of 654. The minimum numbers for this taxon might range through 619 different values, with the actual calculated value depending on how the faunal material was aggregated prior to minimum number determination.

The changes in absolute minimum number abundances that result from different approaches to aggregation are in themselves troublesome because

TABLE 2.6
Maximum Possible Differences in Minimum Number Values, Connley Cave No. 4 Mammals

Taxon	MNI_{site}	NISP	Maximum possible MNI difference
1	36	654	618
2	10	119	109
3	9	77	68
4	8	55	47
5	13	43	30
6	2	27	25
7	5	22	17
8	2	21	19
9	4	13	9
10	1	12	11
11	4	7	3
12	2	5	3
13	1	5	4
14	1	4	3
15	1	4	3
16	3	4	1
17	1	3	2
18	2	2	0
19	1	1	0
20	1	1	0
21	1	1	0
22	1	1	0

they may greatly change the magnitude of the numbers with which the analyst is working. As discussed above, even if the ratios of taxonomic abundances among taxa remain the same when larger aggregation units are used, significance tests applied to the smaller numbers obtained from these larger units will give very different exact probabilities compared with those obtained when more divisive approaches are employed. As a result, the meaning of significance tests applied to minimum numbers becomes clouded (see also Cowgill 1977).

More serious is the fact that the distributions of most abundant elements will almost always be such as to cause different aggregation methods to differentially alter the absolute abundances of taxa as measured by minimum numbers. This will fail to occur only when the most abundant elements of all taxa are identically distributed across aggregation units. At Connley Cave No. 4, this is not the case, as may be seen from Table 2.7, which shows the distribution of most abundant elements for two taxa across levels in Stratum 3 of this site. Most abundant elements are not identically distributed in this instance; with the

TABLE 2.7

Distribution of Most Abundant Elements within Stratum 3 for Two Connley Caves No. 4 Mammals

Stratum 3, level	Taxon 4	Taxon 5
24	1 right mandible	—
25	—	—
26	6 right mandible	4 right innominate
27	2 right mandible	3 right femur
28	—	1 left innominate
29	2 right mandible	1 right innominate
30	2 left mandible	1 right innominate
31	1 maxilla	—
32	1 right innominate	1 left mandible

exception of those taxa for which there are very few identified specimens, this is the case for all other Connley Cave No. 4 taxa as well. Accordingly, it is to be expected that different aggregation methods will differentially alter taxonomic abundances based on minimum numbers of individuals at this site.

Table 2.8 shows these expected results for selected pairs of the five taxa represented by the greatest numbers of identified specimens at Connley Cave No. 4. No ratios of abundance are identical across aggregation methods, and many are widely disparate. Expectations based on considerations of the distribution of most abundant elements across aggregation units are fully met. At Connley Cave No. 4, little faith can be placed in the ratios of taxonomic abundance indicated by minimum numbers. The same is true for any fauna for which most abundant elements are not identically distributed across aggregation units.

Finally, it is important to note that the differentially altered absolute abun-

TABLE 2.8

The Effects of Aggregation on Abundance Ratios Based on Minimum Numbers: Selected Pairs of Connley Cave No. 4 Mammals

Taxon pair	Abundance ratios		
	MNI_{10cm}	$MNI_{stratum}$	MNI_{site}
1–2	1.89	2.60	3.60
2–3	1.65	1.88	1.11
3–4	1.00	0.62	1.13
4–5	1.00	0.93	0.62

dances caused by the effects of differing aggregation methods on minimum numbers may greatly alter the outcome of any significance test applied to minimum number data. Assume, for instance, that an analyst is interested in knowing whether the abundance of *Lepus* spp. changed significantly, compared with the abundance of all other mammals, between Strata 3 and 4 within Connley Cave No. 4. The choice of MNI_{10cm} as the measure of abundance would provide a χ^2 value of .01, and would lead to the conclusion that the relative abundance of this genus did not change significantly between these two strata. If $MNI_{stratum}$ were chosen as the abundance measure, a χ^2 value of 3.58 would result, and, depending on the significance level chosen, might or might not lead to the conclusion that the change in abundance was significant. If NISP (the maximum possible minimum number value) were chosen, a χ^2 value of 49.64 would result, and the conclusion that the change was highly significant would necessarily follow (see Table 2.9).

Hidden Cave The Connley Cave No. 4 mammals illustrate many of the effects that a general consideration of the potential consequences of aggregation leads us to expect. Hidden Cave illustrates other aspects of these effects.

One of the goals of the research at Hidden Cave was to shed light on the nature of environmental change in the southern Lahontan Basin of western Nevada during the past 15,000 years or so. Part of this effort involved the analysis of pollen and plant macrofossils from the sediments of Hidden Cave by P. Wigand and P. J. Mehringer, Jr., who found that the Hidden Cave plant remains reflected two major environmental changes that had occurred in the area since the late Pleistocene (Wigand and Mehringer 1984). First, Wigand and Mehringer demonstrated a sharp decline in pine *(Pinus)* and sagebrush *(Artemisia)* pollen at the top of Stratum XI, a decline that corresponds with the shrinking of Pleistocene lakes and with the replacement of woodland or steppe by shrub-dominated desert vegetation. Second, they noted that *Artemisia* pollen remained relatively abundant until at least 6500 B.P., corresponding roughly with the top of Stratum VI. After this time, *Artemisia* pollen declined in frequency once again, perhaps reflecting the onset or warmer and/or drier conditions and the establishment of vegetation resembling that of modern times. These changes are in line with climatic records from other parts of the arid western United States (Grayson 1982; Mehringer 1977; Spaulding *et al.* 1983; Van Devender and Spaulding 1979).

Of the many kinds of analyses that might be done with the Hidden Cave mammals, one might address whether or not those mammals reflect the changes detected by Wigand and Mehringer. Because their analysis defines three groups of strata that are internally homogeneous as regards floral content (Strata I-V, VI-X, and XI-XIV, corresponding to the late and middle Holocene, early Holocene, and late Pleistocene, respectively), the faunal contents

TABLE 2.9

Comparing Taxonomic Abundances between Strata by Analytic Approach: Connley Cave No. 4 *Lepus* spp[a,b]

Stratum	MNI_{10cm}		$MNI_{stratum}$		NISP	
	A	B	A	B	A	B
3	25	53	14	48	256	247
4	37	76	23	37	388	150

[a] A = *Lepus* spp.; B = all other mammals.
[b] $MNI_{10cm}: \chi^2 = 0.01$, $p > .90$; $MNI_{stratum}: \chi^2 = 3.58$, $.10 > p > .05$; NISP: $\chi^2 = 49.64$, $p < .001$.

of these stratigraphic groups can be compared to one another to see if the abundances of the Hidden Cave mammals changed in concert with the Hidden Cave flora. For instance, the floral records suggest that the relative abundances of *Sylvilagus* (rabbits) and *Lepus* (hares) should differ significantly between Strata I–V, on the one hand, and Strata VI–X, on the other. Individuals of *Sylvilagus* prefer shrubbier habitats than do individuals of *Lepus;* Strata VI–X, therefore, should be characterized by relatively greater numbers of *Sylvilagus* than of *Lepus* compared to Strata I–V.

Judged on the basis of numbers of identified specimens per taxon, this predicted shift occurs: *Sylvilagus* is significantly more abundant than *Lepus* in Strata VI–X and significantly less abundant in Strata I–V (171 specimens of *Sylvilagus* and 89 of *Lepus* in Strata VI–X, compared with 620 specimens of *Sylvilagus* and 956 of *Lepus* in Strata I–V; $\chi^2 = 53.35$, $p < .001$). The problem of interdependence suggests, however, that such an analysis is not properly done using specimen counts. Thus, the apparently impressive results obtained from *Sylvilagus* and *Lepus* specimen counts must be set aside (and the results are only apparently impressive, for reasons that will be discussed in Chapter 3). Instead, minimum numbers must be calculated, since minimum numbers for any given aggregation unit must be independent of one another. In the Hidden Cave setting, there are two obvious approaches available for aggregation. Each of the 14 strata can be treated as providing a separate faunal aggregate, minimum numbers defined 14 times per taxon, and those separate numbers then added to provide values for Strata I–V, VI–X, and XI–XIV. Alternatively, because we are interested in comparing the faunas of these three groups of strata, each group could be treated as providing a separate faunal aggregate, minimum numbers determined three times for each taxon, and those numbers form the basis for comparison. The results of both approaches to minimum number determination are shown in Table 2.10.

TABLE 2.10

Minimum Numbers of Individuals for the Hidden Cave Mammals, Calculated by Separate and by Grouped Strata

	Strata as separate aggregates				Strata as lumped aggregates			
	I–V	VI–X	XI–XIV	Σ	I–V	VI–X	XI–XIV	Σ
Sorex palustris	1	—	—	1	1	—	—	1
Myotis sp.	1	—	—	1	1	—	—	1
Myotis yumanensis	4	1	—	5	4	1	—	5
Antrozous pallidus	2	—	—	2	2	—	—	2
Sylvilagus sp.	8	3	1	12	3	2	1	6
S. cf. nuttallii	28	10	3	41	19	8	1	28
S. nuttallii	9	2	1	12	7	1	1	9
Lepus sp.	32	5	3	40	21	3	1	25
Lepus californicus	3	—	—	3	3	—	—	3
Marmota flaviventris	12	3	1	16	8	2	1	11
Ammospermophilus cf. leucurus	1	1	1	3	1	1	1	3
A. leucurus	3	—	—	3	2	—	—	2
Spermophilus sp.	6	2	1	9	3	1	1	5
S. cf. townsendii	3	1	—	4	3	1	—	4
S. townsendii	8	1	1	10	6	1	1	8
Thomomys sp.	5	3	1	9	2	3	1	6
T. cf. bottae	4	—	—	4	3	—	—	3
T. bottae	4	1	—	5	3	1	—	4
Perognathus sp.	7	4	3	14	5	4	3	12
P. longimembris	6	1	—	7	6	1	—	7
P. formosus	8	3	1	12	7	2	1	10
P. parvus	3	1	1	5	3	1	1	5
Microdipodops sp.	1	—	—	1	1	—	—	1
Dipodomys sp.	37	10	1	48	31	9	1	41
D. ordii	1	—	—	1	1	—	—	1
D. cf. microps	4	—	—	4	2	—	—	2
D. microps	13	1	—	14	9	1	—	10
D. cf. deserti	3	—	—	3	3	—	—	3
D. deserti	1	—	—	1	1	—	—	1
Reithrodontomys megalotis	2	—	—	2	2	—	—	2
Peromyscus sp.	6	1	3	10	6	1	2	9
P. maniculatus	7	2	1	10	6	2	1	9
P. crinitus	1	1	—	2	1	1	—	2
Onychomys sp.	—	1	—	1	—	1	—	1
Neotoma sp.	11	4	1	16	10	3	1	14
N. cf. lepida	17	4	4	25	11	3	2	16
N. lepida	21	4	3	28	15	3	2	20
N. cf. cinerea	17	6	2	25	13	4	2	19
N. cinerea	20	7	1	28	15	7	2	24
Microtus sp.	35	3	—	38	30	3	—	33

TABLE 2.10 (Continued)

	Strata as separate aggregates				Strata as lumped aggregates			
	I–V	VI–X	XI–XIV	Σ	I–V	VI–X	XI–XIV	Σ
M. montanus	7	1	—	8	6	1	—	7
Ondatra zibethicus	3	—	—	3	2	—	—	2
Canis cf. latrans	3	—	—	3	1	—	—	1
C. latrans	5	1	—	6	2	1	—	3
C. lupus	2	—	—	2	1	—	—	1
Vulpes vulpes	3	1	1	5	1	1	1	3
Martes sp.	—	1	—	1	—	1	—	1
Martes nobilis	1	—	—	1	1	—	—	1
Mustela sp.	1	—	—	1	1	—	—	1
M. cf. frenata	4	1	1	6	1	1	1	3
M. frenata	2	1	—	3	1	1	—	2
M. vison	2	—	—	2	1	—	—	1
Taxidea taxus	2	1	—	3	1	1	—	2
Spilogale putorius	2	1	—	3	1	1	—	2
Mephitis mephitis	1	—	—	1	1	—	—	1
Lynx cf. rufus	2	—	—	2	1	—	—	1
Equus sp.	2	—	1	3	1	—	1	2
Camelops cf. hesternus	1	—	—	1	1	—	—	1
Odocoileus cf. hemionus	1	—	—	1	1	—	—	1
Antilocapra americana	1	—	1	2	1	—	1	2
	400	94	38	532	296	79	31	406

Table 2.10 shows many of the same effects displayed by the Connley Cave No. 4 mammals. The great reduction in sample size (3877 identified specimens, compared to 532 or 406 individuals, a reduction of 86% and 90%, respectively), due to the elimination of all but the most abundant element for each taxon in each aggregate, is obvious. Differential changes in relative taxonomic abundances between the two sets of minimum numbers can also be readily found by scanning the table. Such alterations are once again due to the selection of only most abundant elements.

The relative abundances of *Sylvilagus* and *Lepus* illustrate some of these changes. The differences in the relative abundances of these taxa between Strata I–V on the one hand, and VI–X on the other, are no longer significant, whether the assessment is made using minimum numbers calculated by separate strata ($\chi^2 = 2.34$, $p > .20$) or by combined strata ($\chi^2 = 2.62$, $p > .20$).

Part of this difference is due solely to differences in sample size (Payne 1972a). The absolute abundance of *Lepus* in Strata I–V as calculated using separate strata as aggregation units was reduced 96% from that abundance as calculated from numbers of identified specimens (from 956 to 35). If the minimum number value of *Lepus* in Strata I–V is increased to 956 and all other *Lepus* and *Sylvilagus* separate strata minimum number values increased proportionately, the differences are once again significant ($\chi^2 = 45.90$, $p < .001$). The same shift to significance occurs if similar changes are made for minimum numbers calculated by grouped strata ($\chi^2 = 69.55$, $p < .001$). The differences in these χ^2 values reflect the differential changes in abundances that have occurred as a result of the differential selection of most abundant elements in each of the three approaches to aggregation (see Table 2.11).

Even though Hidden Cave (Table 2.10) displays the same kinds of aggregation effects as displayed by the Connley Cave No. 4 mammals (Table 2.5), it is also true that these effects are not as pronounced for the Hidden Cave mammals as they are for the Connley Cave No. 4 mammals. The reason is simple. For the Connley Cave No. 4 mammals, MNI_{10cm} was calculated on the basis of 34 separate aggregates, each requiring a separate definition of most abundant element for each taxon present. As a result, the differences between the absolute abundances provided by the most divisive (MNI_{10cm}) and the most agglomerative (MNI_{site}) approaches to aggregation were pronounced. I have presented two approaches to aggregation for the Hidden Cave mammals, neither of which is as divisive as MNI_{10cm} or as agglomerative as MNI_{site} for Connley Cave No. 4. Thus, the two sets of Hidden Cave values differ less. In addition, the Hidden Cave vertebrates are very unevenly distributed. Within Strata XI–XIV, for instance, 91% of the identified specimens are within Stratum XIII; within Strata VI–X, 73% of the identified specimens are within Stratum VII; within Strata I–V, 44% of the identified specimens are within Stratum IV and an additional 34% within Stratum V (see Table 1.3). Because the identified specimens are so unevenly distributed across strata, and because so many taxa are represented by very few specimens, the effects of different approaches to aggregation are less pronounced than they would otherwise be. Within Strata XI–XIV, for example, treating all faunal material as a single aggregate provides minimum numbers almost identical to those that result from aggregating that material by separate strata because 91% of the identified specimens come from a single stratum (Stratum XIII; see Grayson 1974b for a discussion of this phenomenon in a different setting).

While the effects of aggregation on minimum numbers are less at Hidden Cave than would have been the case had the bones been distributed across strata in a different fashion, it is still true that the choice of aggregation approach here can greatly affect the results of certain kinds of minimum number-based analyses. Analysis of percentage survival of skeletal parts provides a case in point.

TABLE 2.11

Comparisons of Relative Abundance of *Sylvilagus* and *Lepus* in Strata I-V and VI-X, Hidden Cave

	Strata	
	I-V	VI-X
A. NISP[a]		
Sylvilagus	620	171
Lepus	956	99
B. MNI BY SEPARATE STRATA[b]		
Sylvilagus	37	12
Lepus	32	5
C. MNI BY GROUPED STRATA[c]		
Sylvilagus	26	9
Lepus	21	3
D. MNI BY SEPARATE STRATA[d,e]		
Sylvilagus	1105	359
Lepus	956	149
E. MNI BY GROUPED STRATA[d,f]		
Sylvilagus	1184	410
Lepus	956	137

[a] $\chi^2 = 53.35$, $p < .001$.
[b] $\chi^2 = 1.60$, $p > .20$.
[c] $\chi^2 = 1.54$, $p > .20$.
[d] Values for *Lepus*, Strata I-V, set at NISP value for *Lepus*, Strata I-V; all other MNI values changed proportionately.
[e] $\chi^2 = 45.90$, $p < .001$.
[f] $\chi^2 = 69.55$, $p < .001$.

Studies of percentage survival of skeletal parts have been used in a variety of ways in a variety of studies, including a classic and extremely valuable set of analyses of the Makapansgat fauna by Brain (1969, 1981). Nearly all these studies begin by calculating the minimum number of individuals per taxon for a given faunal aggregate (the exceptions are provided by Binford and his followers; Binford's approach is discussed at the end of this chapter). The number of elements expected on the basis of these minimum number values are then compared to the number of those elements actually observed. Table 2.12 presents a simple example. The fauna depicted here provided 48 specimens of Taxon A, distributed across elements as shown in Table 2.12A. A minimum of 10 individuals for this taxon is indicated by the presence of 10 right mandibles.

TABLE 2.12
The Calculation of Percentage Survival of Skeletal Parts: An Example[a]

A. Number of identified specimens by element, Taxon A

Element	NISP
RMD	10
LMD	8
RH	9
LH	1
RU	1
LU	4
RF	5
LF	5
RT	3
LT	2

B. Determination of Percentage Survival, Taxon A skeletal parts

Element	"Expected"	Observed	% Survival
MD	20	18	90
H	20	10	50
U	20	5	25
F	20	10	50
T	20	5	25

[a] D, distal; F, femur; H, humerus; L, left; MD, mandible; R, right; T, tibia; U, ulna.

Assuming that entire skeletons had been initially deposited (an assumption implied by the term *percentage survival* for the values being calculated; a different facilitating assumption would allow the target of analysis to become the *percentage originally deposited*, although neither assumption is usually warranted), then 10 left mandibles must also have been present at one time, as well as 20 humeri, 20 ulnae, and so on. Since there are only 18 mandibles of this taxon in the fauna, only 90% have "survived" (see Table 2.12B).

How might aggregation affect percentage survival values? That the effect might be pronounced is probably already clear, but a simple example can be drawn from Hidden Cave. I will present two calculations of percentage survival of skeletal elements of *Lepus* from Strata I – V at this site. Rather than examining percentage survival of all skeletal elements, I will focus on those elements that defined minimum numbers in one or more cases: the ulna, humerus, tibia, and second, third, and fourth metatarsals.

The first of my calculations of percentage survival is based on the minimum

TABLE 2.13

The Calculation of Percentage Survival of Skeletal Parts: Hidden Cave *Lepus*, Strata I–V, Using Minimum Numbers Calculated on a Single Stratum Basis[a]

Element	Expected	Observed	Element	Expected	Observed
STRATUM I: MNI = 4 (RDH)			STRATUM II: MNI = 7 (RMT IV)		
PU	8	3	PU	14	7
DU	8	1	DU	14	1
PT	8	1	PT	14	2
DT	8	2	DT	14	3
PH	8	2	PH	14	3
DH	8	4	DH	14	4
MT II	8	2	MT II	14	5
MT III	8	1	MT III	14	6
MT IV	8	5	MT IV	14	8
STRATUM III: MNI = 3 (LDT)			STRATUM IV: MNI = 12 (RDH)		
PU	6	0	PU	24	4
DU	6	1	DU	24	4
PT	6	0	PT	24	3
DT	6	4	DT	24	15
PH	6	0	PH	24	4
DH	6	0	DH	24	14
MT II	6	1	MT II	24	11
MT III	6	1	MT III	24	10
MT IV	6	3	MT IV	24	9
STRATUM V: MNI = 6 (RDU, LMT II, LMT III)					
PU	12	6			
DU	12	2			
PT	12	3			
DT	12	7			
PH	12	1			
DH	12	6			
MT II	12	8			
MT III	12	11			
MT IV	12	7			

[a] D, distal; H, humerus; L, left; MT, metatarsal; P, proximal; R, right; T, tibia, U, ulna.

numbers for *Lepus* calculated on a single stratum basis. The minimum number values, the most abundant element that defined those values, and the expected and observed numbers of elements are shown in Table 2.13. Since I am interested in skeletal completeness in Strata I–V as a whole, I have summed the expected and observed values for all five strata, and calculated percentage

TABLE 2.14

The Effects of Aggregation on the Calculation of Percentage Survival of Skeletal Parts: Hidden Cave *Lepus*, Strata I-V[a]

Element	Expected	Observed	Percentage survival
A. SUMMED SINGLE STRATUM MINIMUM NUMBERS[b]			
PU	64	20	31
DU	64	9	14
PT	64	9	14
DT	64	31	48
PH	64	10	16
DH	64	28	44
MT II	64	27	42
MT III	64	29	45
MT IV	64	32	50
B. STRATA I-IV TREATED AS SINGLE AGGREGATE			
PU	42	20	48
DU	42	9	21
PT	42	9	21
DT	42	31	74
PH	42	10	24
DH	42	28	67
MT II	42	27	64
MT III	42	29	69
MT IV	42	32	76

[a] D, distal; H, humerus; MT, metatarsal; P, proximal; T, tibia; U, ulna.
[b] From Table 2.13.

survival on the basis of those summed values (averaging these values would, of course, provide the same results). The resultant percentage survival values are shown in Table 2.14A.

The second of my calculations uses the minimum numbers for *Lepus* determined by treating all *Lepus* material from Strata I-V as a single aggregate. The minimum number values, the most abundant element that defined that value, and the expected and observed numbers of elements are shown in Table 2.14B, as are the resultant percentage survival values.

The results are instructive. Percentage survival for the distal tibia is 48% when calculated by separate strata and 74% when calculated from the single faunal aggregate. Percentage survival for the distal humerus is 44% when calculated by separate strata and 67% when calculated by single faunal aggregate. No percentage survival values are unaffected. The reason for the difference in percentage survival values is, of course, the fact that different approaches to faunal aggregation have produced different minimum number values, and that different most abundant elements are selected as the basis of minimum number definition as faunal aggregates change. As these variables change, so do the percentage survival figures calculated from them.

It is clear, then, that values for percentage survival of skeletal elements determined on the basis of minimum numbers are heavily dependent on decisions concerning faunal aggregation. As a result, analyses of percentage survival values will be analyses not only of survival of elements through time (granting the facilitating assumption in the first place), but also of analytic decisions concerning aggregation. This fact is of crucial importance when considering the interpretation of percentage survival values. Brain (1981), for example, presents percentage survival values for the Swartkrans Member 2 breccia, yet also notes that "evidence is accumulating that Member 2 is a rather heterogeneous stratigraphic entity, embracing breccias and calcified channel fills of several ages . . . for the time being, the fossils they have yielded are considered as a unit" (1981:239–240). It may be expected that when the Member 2 fauna is subdivided, as Brain feels it will be, and percentage survival values are recalculated on the basis of these new faunal aggregates, the results will be quite different from those calculated on the basis of the Member 2 fauna as a whole.

The Relationship of NISP to MNI

To this point, I have argued that both numbers of identified specimens and minimum numbers of individuals per taxon are associated with difficulties that seem to greatly detract from their utility as measures of taxonomic abundance. On the one hand, specimen counts tally units — identified bones and teeth — whose independence can reasonably be doubted, and whose mechanical interdependence is better assumed than assumed away. All of our counting procedures assume that the units being counted are not mechanically interdependent, so interdependence poses a major problem as regards the use of specimen-based counts. On the other hand, minimum numbers of individuals have values that are in part, and often in large part, determined by the choices made by the analyst concerning how faunal material should be aggregated prior to minimum number calculation. As a result, when an analyst studies minimum number values, that person is studying not only taxonomic abundances, but also the decisions made concerning aggregation.

In short, specimen counts are plagued by the problem of interdependence, but are unaffected by aggregation (for an exception to this statement, however, see my discussion of the approach taken by Binford [1981, 1984], below). Minimum numbers are in part determined by aggregation but, within aggregates, each minimum number is independent of every other. Given these facts, it becomes of interest to explore the relationship between the results provided by these two very different measures of abundance when they are applied to the same sets of faunal material.

Precisely this relationship was, in fact, explored by Pierre Ducos in 1968, in a very valuable discussion of the quantification of taxonomic abundances in archaeological faunas (Ducos 1968). Ducos' analysis has been neglected in the English-language literature, as can be seen from the fact that later, identical analyses of the relationship between specimen counts and minimum numbers published in English did not build on Ducos' work, but instead represent independent derivations of the same approach (e.g., Casteel 1976/1977; Hesse 1982). Ducos' analysis is insightful and bears repetition here.

Ducos conducted his analysis because he was interested in calculating the relative abundances of taxa represented in faunas from a series of sites in Palestine. He was critical of minimum numbers as a counting unit because of the great reduction in sample size that occurs when specimen counts are transformed into minimum numbers. Statistical analyses, Ducos observed, often require large samples, and minimum numbers rarely provide such samples even when they are based on sizeable specimen counts. In addition, Ducos noted that minimum number values depend on the particular element chosen to define them; later information, as might be provided by additional excavations, might change those numbers significantly. He also pointed out that, while large samples were important for statistical reasons, the absolute numbers themselves were rarely the target of interest, but were instead used to determine that target: relative abundances. What was needed, he suggested, was a counting unit that would provide relative abundances that approximated those in the faunal population of the site as a whole, and, he argued, specimen counts, not minimum numbers, provided that unit.

Although I do not agree with all the criticisms that Ducos leveled at minimum numbers, the next step he took was an important one. In order to demonstrate the general relationship between specimen counts and minimum numbers, Ducos plotted the logarithms of the number of identified specimens against the logarithms of the minimum number of individuals defined from those specimens, drawing his data from a series of published faunas. He discovered that this relationship was linear. Because untransformed minimum number values increased at decreasing rates as untransformed specimen counts increased, Ducos argued that the abundance of rare species would be overestimated were minimum numbers used to calculate taxonomic frequencies. For these reasons, and for those discussed in the preceeding paragraph, Ducos rejected minimum numbers and based all of his calculations of taxonomic abundance on specimen counts.

Nearly a decade later, Casteel (1976/1977) replicated Ducos' results. Al-

THE RELATIONSHIP OF NISP TO MNI 51

though Casteel's prime interest at this time was the relationship between the variables MNI/NISP and NISP (see below), he also examined the relationship between MNI and NISP. His sample for this examination included 610 data points that had been drawn from a wide variety of archaeological and paleontological sites. Casteel found that the relationship between minimum numbers and specimen counts in this sample was curvilinear for faunas that contained less than 1000 specimens and linear for those of larger size (MNI = $0.78(NISP)^{0.52}$ and MNI = $5.56 + 0.0225(NISP)$, respectively).

Casteel derived these results by arbitrarily dividing his samples at 1000 identified specimens, and apparently did not examine the relationship between minimum numbers and specimen counts across all values in his data set. As a result, it is not possible to know from his paper if his second, linear relationship might represent very slow change in minimum numbers at high specimen counts, such as might be accounted for by a power function fit to his entire data set, or if the two relationships actually differed in that data set. That the apparently linear relationship did not result from the relationship MNI = $0.78(NISP)^{0.52}$ is shown by the fact that by the time NISP reaches 5000, Casteel's power function predicts an MNI of 65.40, while his linear function predicts an MNI of 118.06; at NISP = 10,000, the corresponding MNI values are 93.78 and 230.56. Since Casteel presented only the best fit lines for these relationships and did not plot the data points around those lines, it is not possible to visually examine the residuals to examine this question. Since he did not publish a list of the faunas he had used, his relationship cannot be examined in more detail now. However, it would seem that his results generally conform with those obtained by Ducos, whose curves had been fit by inspection.

Both Ducos and Casteel examined the relationship between MNI and NISP in composite samples in which taxa from many different sites were combined in a single analysis. Although the use of composite samples of this sort is essential if the goal of analysis is to derive a single equation that will best allow estimation of MNI values from a given set of NISP values for a site not included in the composite, as Casteel (1976/1977) was attempting to do, this approach allows the introduction of factors that may obscure the relationship between MNI and NISP within any given fauna. If, for instance, the composite sample incorporates minimum number values calculated by different approaches to aggregation or according to different operational definitions of minimum numbers, the relationship between MNI and NISP across all sites may not apply to any single site in the composite. Detailed analysis of residuals as part of the regression procedure might detect this phenomenon, but such analyses have not been conducted in the studies published to date.

Examining the relationship between MNI and NISP on a single site basis avoids these problems. What form do these relationships take? The faunas I have examined strongly suggest that the curvilinear relationships posited by

TABLE 2.15
Minimum Numbers of Individuals and Numbers of Identified Specimens per Taxon, Prolonged Drift, Kenya[a]

Taxon	NISP	MNI
Domestic cattle	250	22
Kongoni (hartebeest)	232	18
Wildebeest	241	17
Common zebra	464	16
Thomson's gazelle	165	16
Grant's gazelle	106	11
Impala	35	7
Domestic caprine	50	5
Eland	37	4
Buffalo	11	2
Giraffe	6	1
Warthog	14	1
White rhinoceros	4	1
Leporid	1	1
Mole Rat	28	7
	1644	129

[a] From Gifford et al. 1980.

Ducos and Casteel are, in fact, general ones. The fauna from the Pastoral Neolithic site of Prolonged Drift (ca. 2500 B.P.), near Lake Nakuru in southwestern Kenya, illustrates the relationship well. Gifford et al. (1980) identified 15 mammalian taxa from Prolonged Drift from a total of 1644 identified specimens (Table 2.15). Figure 2.3 illustrates the relationship between untransformed numbers of identified specimens and untransformed minimum numbers of individuals. A runs test (Draper and Smith 1966) on the residuals plotted against NISP values shows that the sequence of positive and negative residuals (that is, deviations above and below the regression line) is significant ($p < .043$), indicating that a linear model inappropriately describes this relationship. Figure 2.4 plots the relationship between $\log_{10}$ NISP and $\log_{10}$ MNI; the residuals that result from this relationship indicate that this fit is an appropriate one. The regression equation between the MNI and NISP values for Prolonged Drift is MNI = $0.49(\text{NISP})^{.64}$, similar to that obtained by Casteel (1976/1977) for his composite sample of taxa represented by fewer than 1000 specimens.

Other faunas provide similar results. Five examples are shown in Figures 2.5 through 2.9, illustrating the relationship between $\log_{10}$ MNI and $\log_{10}$ NISP in a wide variety of faunas. In all cases, analyses of residuals show that untransformed MNI and NISP values are not related in a linear fashion, but that MNI and

THE RELATIONSHIP OF NISP TO MNI

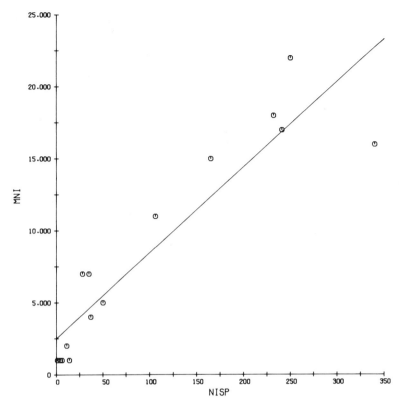

Figure 2.3 The relationship between MNI and NISP, Prolonged Drift, Kenya (Gifford *et al.* 1980).

NISP transformed by common logarithms are. Table 2.16 summarizes the regression equations for the best fit lines, and the correlation coefficients, for these five sites and for Prolonged Drift.

These results fully support Ducos' conclusion that the relationship between MNI and NISP is log–log linear. In addition, I note that functions of the form MNI = a(NISP)^b describe the relationship for faunas that consist of well in excess of 1000 identified specimens (see, for instance, Figures 2.5 and 2.6). I suspect that Casteel (1976/1977) concluded that two different sorts of relationship pertained to faunas represented by less than, and more than, 1000 specimens simply because he divided his sample at this point and analyzed each subset separately. He either had too few data points to recognize that a linear fit was inappropriate for his large samples, or did not analyze the residuals in order to see if at very high samples sizes, individuals were being added at a decreasing rate. As I have noted, this question cannot be answered without access to his sample.

TABLE 2.16

Regression Equations and Correlation Coefficients for the Relationship Between MNI and NISP at Prolonged Drift (Figure 2.4), Apple Creek (Figure 2.5), Buffalo (Figure 2.6), Dirty Shame Rockshelter Stratum 2 (Figure 2.7), Dirty Shame Rockshelter Stratum 4 (Figure 2.8), and Fort Ligonier (Figure 2.9)

Site	Regression equation	r	p
Prolonged Drift	MNI = .49(NISP)$^{.64}$	.939	<.001
Apple Creek	MNI = .69(NISP)$^{.59}$	.937	<.001
Buffalo	MNI = .71(NISP)$^{.63}$	.957	<.001
Dirty Shame Stratum 2	MNI = 1.03(NISP)$^{.64}$	.971	<.001
Dirty Shame Stratum 4	MNI = 1.04(NISP)$^{.62}$	.978	<.001
Fort Ligonier	MNI = 1.11(NISP)$^{.40}$	.785	<.001

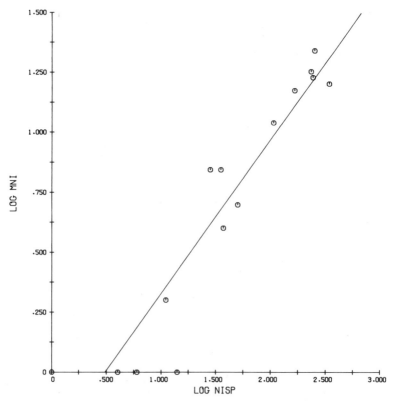

Figure 2.4 The relationship between $\log_{10}$ MNI and $\log_{10}$ NISP, Prolonged Drift, Kenya.

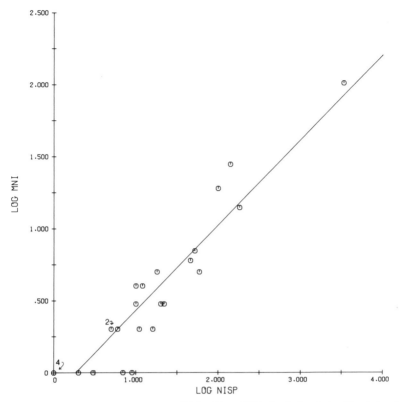

Figure 2.5 The relationship between $\log_{10}$ MNI and $\log_{10}$ NISP, Apple Creek site, Illinois (ca. 250 B.C. to A.D. 750): 4110 specimens of 26 mammalian taxa from the midden and plow zones (Parmalee et al. 1972).

It would seem, then, that within many vertebrate faunas the minimum number of individuals and the number of identified specimens per taxon are related in a curvilinear fashion, the relationships being well described by an equation of the form MNI = $a(\text{NISP})^b$, where a and b are constants that must be empirically derived in each instance. Hesse (1982), however, has examined the relationship between MNI and NISP in four faunas, and has argued that this relationship is linear. It is instructive to examine Hesse's results in detail.

One of Hesse's faunas is a composite drawn from Parmalee's analysis of a number of avian faunas from a series of sites in the Plains region of North America, all dating to between about A.D. 900 and 1800. The very linear plot provided by Hesse (1982) from this sample is somewhat difficult to interpret. The data presented by Parmalee (1977) represent a composite not only because 51 sites are represented, but also because the data that Parmalee pro-

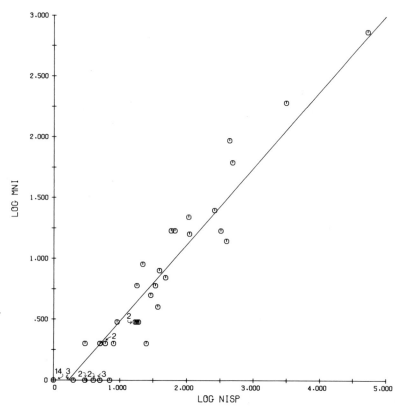

Figure 2.6 The relationship between $\log_{10}$ MNI and $\log_{10}$ NISP, Buffalo site, West Virginia (ca. A.D. 1650): 58,689 specimens of 54 taxa of birds and mammals (Guilday 1971).

vides are taxonomic composites. Parmalee (1977) was interested in examining the relative abundances of bird families in these 51 sites, and the MNI and NISP values he presents were accordingly amassed by summing all the MNI and NISP values for a given avian family from all 51 sites. Thus, his figures for the Anatidae (459 identified specimens from a minimum of 130 individuals) represent the summed values for 13 species of anatids from all 51 sites. These were the figures Hesse (1982) plotted to obtain his linear relationship between MNI and NISP: composites merging values from many species and many sites that were perfectly appropriate for Parmalee's purposes, but that introduce many complexities as regards understanding the relationship between MNI and NISP within faunal assemblages. More confusing, however, is the fact that Hesse's plot of Parmalee's data (Hesse 1982: Figure 26) purports to show the relationship between untransformed MNI and NISP values (hence his discussion of the linearity of the relationship displayed). In this plot, the displayed NISP values go

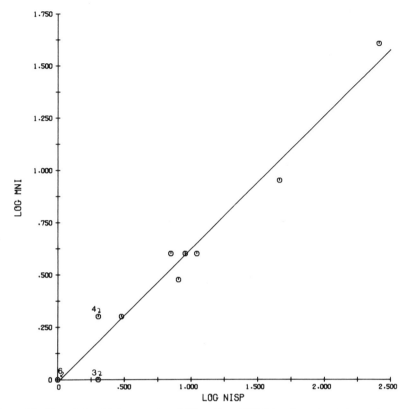

Figure 2.7 The relationship between $\log_{10}$ MNI and $\log_{10}$ NISP, Dirty Shame Rockshelter, Oregon, Stratum 2 (1100–2700 B.P.): 363 specimens of 20 mammalian taxa (Grayson 1977b).

no higher than 8 (Parmalee's table contains four NISP values of greater than 100), and the displayed MNI values go no higher than 6 (Parmalee's table contains three MNI values of greater than 100). Hesse's plot of Parmalee's data can be duplicated only by performing $\log_e$ tranforms of both MNI and NISP values. Thus, far from being linear, the relationship between MNI and NISP in Parmalee's composite sample is curvilinear. That relationship is displayed in Figure 2.10, which is also identical to Hesse's Figure 26; the equation describing the best-fit line for the relationship between MNI and NISP for Parmalee's data is MNI = 0.75 (NISP)$^{.83}$ ($r = 0.983$, $p < .001$).

Hesse's remaining samples are drawn from unpublished dissertations, and I am unable to reanalyze them. These do not deal with entire faunas, however, but with caprine remains only. Hesse suggests that the relationship between MNI and NISP in these caprine faunas is linear. Certainly, linear relationships of this sort exist, the classic example being provided by certain invertebrate faunas

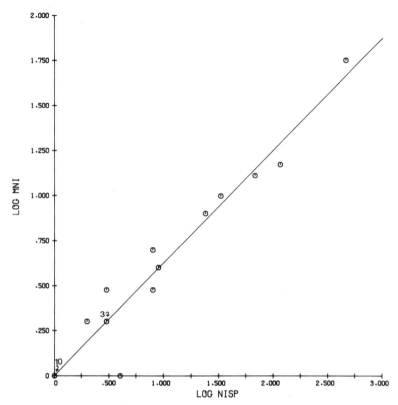

Figure 2.8 The relationship between $\log_{10}$ MNI and $\log_{10}$ NISP, Dirty Shame Rockshelter, Oregon, Stratum 4 (6300–6800 B.P.): 757 specimens of 24 mammalian taxa (Grayson 1977b).

in which each identified specimen contributes an individual, a situation that frequently occurs with gastropods. Small mammal faunas also provide examples of such linear relationships, since small mammals are often identified on the basis of very restricted numbers of elements. John Guilday's analyses of late Pleistocene and Holocene faunas from eastern North America provide many examples. Figure 2.11, for instance, presents the relationship between MNI and NISP for five species of shrews from the late Pleistocene and Holocene deposits of Clark's Cave, northwestern Virginia (Guilday *et al.* 1977). This relationship is markedly linear (MNI = 0.67 + 0.52(NISP), $r = .999$, $p < .001$), and thus unlike the situations I have discussed above. The reason is that only a small number of elements are used to identify such animals, and hence to define minimum numbers for them. As a result, the chances of drawing a given most abundant element from a faunal sample remain stable across taxa in the case of Guilday's shrews, or across strata (apparently) in the case of Hesse's caprines.

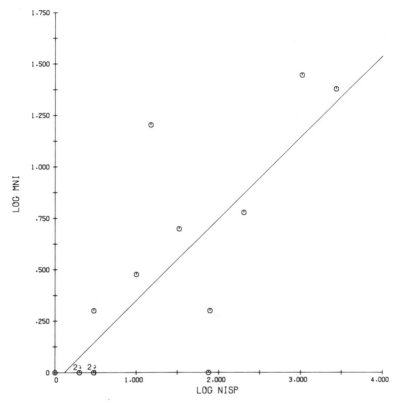

Figure 2.9 The relationship between $\log_{10}$ MNI and $\log_{10}$ NISP, Fort Ligonier, Pennsylvania (A.D. 1758–1766): 4497 specimens of 14 mammalian taxa from a historic British relay station (Guilday 1970).

Table 2.17, for instance, presents the number of identified specimens per shrew species at Clark's Cave. Guilday identified only skulls and mandibles to the species level, and in only one case did Clark's Cave contain even a single shrew skull that could be identified to species. In essence, there were only two elements that could define minimum numbers of individuals for these species: right and left mandibles. The application of binomial tests shows that in no case does the number of right mandibles differ significantly from the number of left mandibles for any species of Clark's Cave shrew. The Clark's Cave shrew mandibles could readily have been obtained by randomly sampling a population of shrew mandibles that contained an equal number of rights and lefts (indeed, there is no reason to think that such a population was not being sampled here). As a result, the relationship between numbers of identified specimens and minimum numbers of individuals derived from those specimens

TABLE 2.17
Identified Specimens of Shrews from Clark's Cave, Virginia[a]

Taxon	Identified specimens	Most abundant element	MNI
Sorex arcticus	1 partial skull 13 left mandibles 12 right mandibles	Left mandible	13
Sorex cinereus	67 left mandibles 61 right mandibles	Left mandible	67
Sorex dispar	2 left mandibles 4 right mandibles	Right mandible	4
Sorex fumeus	9 left mandibles 10 right mandibles	Right mandible	10
Sorex palustris	3 left mandibles 7 right mandibles	Right mandible	7

[a] Guilday et al. 1977.

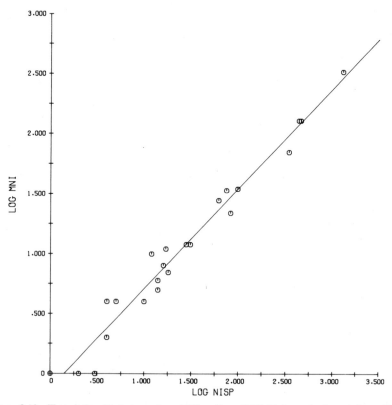

Figure 2.10 The relationship between $\log_{10}$ MNI and $\log_{10}$ NISP, bird remains from Arikara village sites (ca. A.D. 900–1800) (Parmalee 1977).

THE RELATIONSHIP OF NISP TO MNI

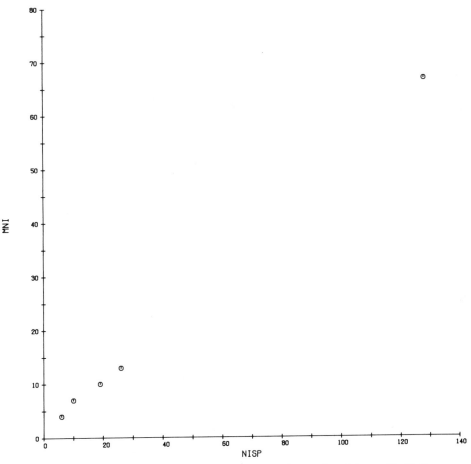

Figure 2.11 The relationship between MNI and NISP for five species of shrews, Clark's Cave, Virginia (Guilday *et al.* 1977).

is linear across the five species of Clark's Cave shrews, and the slope of that relationship is almost exactly 0.50.

In short, while the relationship between MNI and NISP is curvilinear in most vertebrate faunas I have examined, it need not be in all, and Hesse (1982) is certainly correct in concluding that there are faunas in which it is linear. In general, the slope of the relationship between MNI and NISP within any given faunal collection will be a function of the probability of drawing a most abundant element across all aggregation units and across all taxa, as the faunal assemblages of those aggregates and taxa are sampled without replacement. While this relationship is commonly curvilinear, it may be linear for subsets of those faunas, as with Guilday's shrews. It can even be linear for entire faunas, as with gastropod collections, or with collections composed entirely of small mammals, in which only mandibles or complete skulls allow the definition of an individual for each species involved. In most cases, however, as the sample size for a given taxon increases, the chances of drawing a most abundant element decrease, and the relationship between MNI and NISP is curvilinear within the assemblage as a whole. Because of this fact, Casteel's attempt to derive a single equation from a composite sample that could be used to obtain MNI values from NISP counts for any vertebrate fauna was premature. He was, however, correct in stressing the predictable relationship between MNI and NISP in any given fauna, and it is this point I wish to stress here. *For any given fauna, MNI values can normally be tightly predicted from NISP counts* (see the coeffi-

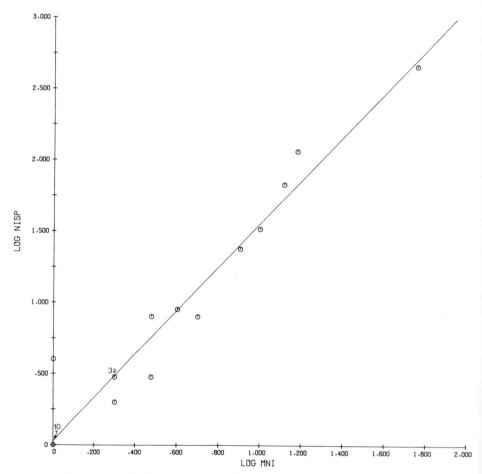

Figure 2.12 The relationship between $\log_{10}$ NISP and $\log_{10}$ MNI for the Dirty Shame Rockshelter Stratum 4 mammals.

TABLE 2.18
Regression Equations and Correlation Coefficients for the Relationship between NISP and MNI at Prolonged Drift, Apple Creek, Buffalo, Dirty Shame Rockshelter Stratum 2, Dirty Shame Rockshelter Stratum 4, and Fort Ligonier

Site	Regression equation	r	p
Prolonged Drift	NISP = $4.16(MNI)^{1.37}$	.939	<.001
Apple Creek	NISP = $2.36(MNI)^{1.49}$	.937	<.001
Buffalo	NISP = $2.00(MNI)^{1.45}$	.957	<.001
Dirty Shame Stratum 2	NISP = $1.12(MNI)^{1.48}$	.971	<.001
Dirty Shame Stratum 4	NISP = $1.07(MNI)^{1.53}$	.978	<.001
Fort Ligonier	NISP = $3.74(MNI)^{1.55}$	.785	<.001

cients in Table 2.16). As a result, the information on relative abundance that resides in MNI counts generally resides as well in NISP counts, and if relative abundance is the target of analysis, there would seem little reason to spend the time and effort to calculate minimum numbers.

It might be thought that this argument is backwards. After all, the number of identified specimens per taxon really depends on the number of individual animals whose remains were deposited in the site, and it could be argued that this is the reason why one can predict minimum numbers from specimen counts. After all, the regressions I have presented, in which I treat minimum numbers as the dependent variable, can easily be reversed and the results are, of course, equally impressive. One such relationship, for the mammalian fauna of Dirty Shame Rockshelter Stratum 4 (Grayson 1977b), is presented in Figure 2.12, while Table 2.18 presents the regression equations for the best-fit lines that result when NISP is treated as the dependent variable for the six sites illustrated above (Figures 2.4–2.9). Minimum numbers, it might be argued, represent the independent variable, specimen counts the dependent one.

This argument might be powerful were it not for the fact that is assumes that minimum numbers provide an accurate estimate of the actual number of individuals deposited in the site. While it is certainly true that NISP depends on the number of individual animals or parts of animals deposited in a site, it is not true that minimum numbers necessarily measure that variable (see also Damuth 1982; Fieller and Turner 1982; Gilbert and Singer 1982). As I have discussed at length, minimum numbers are heavily dependent on aggregation; they measure, among other things, aggregation as well as relative abundance. This phenomenon is easy to illustrate.

As more and more agglomerative approaches to aggregation are used, it takes more and more specimens to lead to the definition of an individual. As a result, the slope of the relationship between MNI and NISP (the rate at which MNI changes as NISP changes) should reflect aggregation procedures as well as it reflects the chances of drawing a most abundant element in general. Approaches to aggregation that are more agglomerative should be characterized by slopes that are lower than those that characterize more divisive aggregation procedures for the same collection. The Connley Cave No. 4 mammals show this relationship well. The relationship between $\log_{10} \text{MNI}_{10cm}$ and $\log_{10} \text{NISP}$ for the Connley Cave No. 4 mammals is shown in Figure 2.13; the relationship between $\log_{10} \text{MNI}_{stratum}$ and $\log_{10} \text{NISP}$ for these mammals is shown in Figure 2.14. The equations for the best fit lines and the associated correlation coefficients are shown in Table 2.19. Note that the slope for the relationship involving MNI_{10cm} is 0.71; that for the relationship involving $\text{MNI}_{stratum}$ is 0.54. The de-

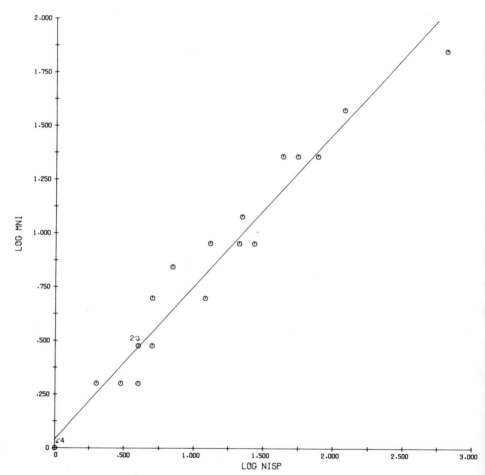

Figure 2.13 The relationship between $\log_{10} \text{MNI}_{10cm}$ and $\log_{10} \text{NISP}$ for the Connley Cave No. 4 mammals.

THE RELATIONSHIP OF NISP TO MNI

TABLE 2.19

Regression Equations and Correlation Coefficients for the Relationship between NISP and MNI_{10cm} and $MNI_{stratum}$ for the Connley Cave No. 4 Mammals

MNI	Regression equation	r	p
10 cm	$MNI = 1.10(NISP)^{.71}$	.982	<.001
Stratum	$MNI = 1.04(NISP)^{.54}$	.916	<.001

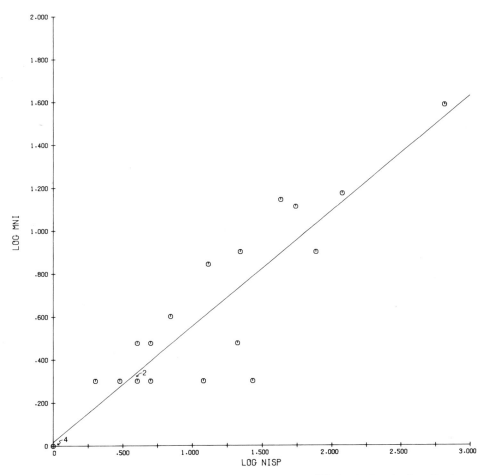

Figure 2.14 The relationship between $\log_{10} MNI_{stratum}$ and $\log_{10}$ NISP for the Connley Cave No. 4 mammals.

crease in slope is measuring aggregation: the more agglomerative approach to aggregation provides a lower slope than the less agglomerative one. For instance, when minimum numbers are calculated for the Connley Cave No. 4 mammals using 10-cm levels, 50 identified specimens leads to the definition of 18 (17.69) individuals; 100 identified specimens leads to the definition of 29 (28.93) individuals. When the fauna is aggregated by strata, 50 specimens define 9 (8.60) individuals, while 100 specimens define 13 (12.50). For MNI_{10cm}, the increase from 50 to 100 specimens leads to the definition of an additional 11 (11.24) individuals, but the same increase in specimen counts leads to the definition of only 4 (3.90) additional individuals when the Connley Cave No. 4 mammals are aggregated by stratum.

Hidden Cave shows the same phenomenon. The equation for the best-fit line describing the relationship between MNI and NISP when the fauna has been aggregated by strata is $MNI = 0.99 (NISP)^{.54}$; when the entire fauna is treated as a single aggregate, the corresponding equation is $MNI = 0.89 (NISP)^{.47}$. Again, the slope decreases as a larger aggregate is used because fewer most abundant elements are defined with larger aggregates, and more specimens are needed to define an individual. Indeed, this effect might explain why the slopes in the three Near Eastern caprine faunas illustrated by Hesse (1982: Figure 27) differ. Hesse suggests the difference may be due to differences in the distribution of most abundant elements within these samples. While this may well be the case, differences in aggregation may also explain the phenomenon he has illustrated.

The slopes of the relationship between MNI and NISP are thus very sensitive to aggregation procedures. It should be clear that minimum numbers measure not only abundance but the effects of aggregation as well. It is for this reason that minimum numbers should not be treated as the independent variable in examining the relationship between MNI and NISP. Proceeding in this fashion would result in treating NISP values not only as a function of numbers of animals (or parts of animals) deposited on the site, which is reasonable, but also as a function of how the fauna happened to be aggregated, which is unreasonable because NISP values are unaffected by aggregation.

I began this examination of the relationship between minimum numbers and the specimen counts from which they are derived by observing that specimen counts are plagued by the problem of interdependence, but are unaffected by aggregation. Several things should now be clear.

First, it should be clear that minimum numbers are necessarily independent of one another only when one can be sure that the faunal aggregates from which they are defined are totally independent of one another. In many cases, total independence can be guaranteed only by treating an entire site as having provided a single faunal aggregate. Certainly, there is no guarantee of independence when arbitrary levels are used as the basis for aggregation, no guarantee when such archaeological features as house floors and storage pits are used, and no guarantee when different midden deposits relating to a single occupation are used. Thus, when Brain (1981) presents his minimum number values, we can be sure that each individual counted is independent of every other individual because he treats each of his cave sites as providing a single faunal aggregate. Guilday's calculation of minimum numbers for the Clark's Cave shrews is also completely free of problems due to specimen interdependence since he treats the entire Clark's Cave bone collection as a single aggregate. However, when minimum numbers are calculated using aggregates whose independence cannot be guaranteed (as with the Connley Cave No. 4 MNI_{10cm} calculations), the prime statistical benefit of minimum numbers has been lost: these numbers are no longer necessarily independent of one another. However, even when we can be sure that the analytic units that form the basis for aggregation are independent units, and that we have therefore avoided the problem of interdependence, we must also be sure that these units could not have been further subdivided. If further subdivision might have been possible, then aggregation effects are likely to be present but hidden from view. While Brain's minimum numbers are necessarily independent of one another, we can be fairly certain that once the stratigraphy of his sites is better understood and subdivision of his faunas becomes possible, the results of all analyses dependent on minimum numbers will be greatly altered. These considerations remove much of the attractiveness of minimum numbers as the basic unit of the quantification of relative taxonomic abundance.

Second, the observation made by Ducos (1968) has been confirmed many times over: minimum numbers are related to specimen counts in a very predictable way. For the faunas that I have examined, and for the composite faunal samples that others have examined, this relationship has been shown to be a simple power function of the form $MNI = a(NISP)^b$. The exceptions are derived from examinations of single taxa: Guilday's shrews and, probably, Hesse's caprines. The exceptions themselves, however, do not suggest that minimum numbers cannot be tightly predicted from specimen counts, but simply involve the precise nature of the equation that relates these two variables. Whatever information on abundances resides in minimum numbers resides as well in specimen counts (the nature of that information will be discussed in Chapter 3).

Given that minimum numbers measure not only abundance but also register the effects of aggregation procedures, it is difficult to see that they have much to offer as a measure of taxonomic abundance. It is, of course, true that minimum numbers can be used as the basis of other kinds of statistical manipulation, such as the extraction of biomass values (e.g., Guthrie 1968), but it should be clear that the effects of aggregation will operate through all measures derived from minimum numbers, as my discussion of the effects of aggregation on the calculation of percentage survival figures illustrates (see also Chapter 6). In short, specimen counts contain the information on relative abundances contained in minimum numbers, but do not register the effects of aggregation and, as Ducos (1968) concluded, there seems no reason not to prefer them to minimum numbers as the basic measure of taxonomic abundance in archaeological faunas. I return to this issue below (see p. 90).

The Relationship of MNI/NISP to NISP

An obvious, but analytically treacherous, measure that can be derived from specimen counts and minimum numbers per taxon is the simple ratio of the two: MNI/NISP, the number of individuals defined per bone for a given taxon, or NISP/MNI, the number of bones per individual for a given taxon. Has one species been more heavily butchered than another? Has a larger proportion of the skeletons of the individuals of one taxon than of another been deposited in a site? MNI/NISP, or its reciprocal, might provide clues to the answers to such questions. Chaplin (1971:67), for instance, suggested that in economically oriented studies, the faunal analyst might "divide the number of fragments by the number of animals represented. A low ratio then suggests butcher's meat, a high ratio, the much fuller use of individual animals." Shotwell (1955, 1958, 1964) used a similar measure to distinguish mammals derived from local and distant communities in paleontological samples, while Thomas (1971) used such a measure to distinguish cultural from natural bones in archaeological sites. Because the ratio appears to be of great analytic value, its use has been widely recommended (e.g., Schiffer 1983; Shipman 1981; Wing and Brown 1979). Unfortunately, it measures sample size.

As I have discussed elsewhere (Grayson 1978a), the relationship between MNI/NISP and NISP must vary between two very easily defined limits. Given that the presence of one identified specimen requires the definition of one individual, and thus an MNI/NISP value of 1.0, it is possible, though very unlikely in vertebrate faunas, that each additional specimen will also define an additional individual. This relationship, shown in Figure 2.15, Line A, is rather frequently encountered in analyses of certain kinds of invertebrates in which only one part of the animal can be identified or in which only entire, or nearly entire, exoskeletons are identified. Gastropods again provide an example (e.g., Guilday et al. 1977). The other limit is provided by situations in which only the first specimen defines an individual, additional identifications not allowing the definition of further individuals (Figure 2.15, Line B). I am aware of no sizeable vertebrate fauna that displays this pattern.

In virtually all vertebrate faunas, the relationship between MNI/NISP and NISP is hyperbolic, of the form MNI/NISP $= a(\text{NISP})^b$, in which a is a positive and b a negative constant that must be derived empirically for any given fauna. That is, as NISP per taxon increases, the value of MNI/NISP decreases at a

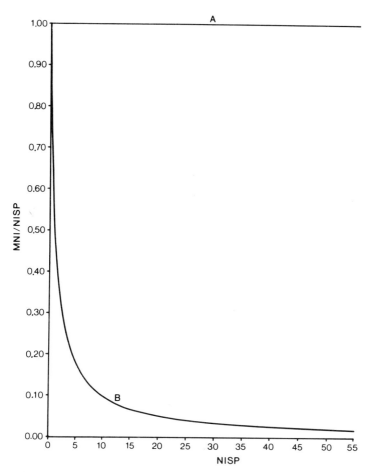

Figure 2.15 The limits between which the relationship of MNI/NISP to NISP must vary: A, each identified specimen defines an individual; B, only the first identified specimen defines an individual.

decreasing rate. This, of course, is no surprise, since NISP appears on both axes of the relationship, once as the independent variable and once as the denominator of the dependent variable. Were the variable of interest NISP/MNI, the appropriate equation becomes, of course, NISP/MNI = $a(NISP)^b$, in which the slope is now positive: NISP/MNI increases at a decreasing rate as NISP increases.

I have illustrated this relationship a number of times elsewhere (Grayson 1978a,b), and present only two examples here. The relationship between $\log_{10}$ MNI/NISP and $\log_{10}$ NISP for the Buffalo Site birds and mammals (Guilday 1971) is shown in Figure 2.16. The equation for the best-fit line describing this

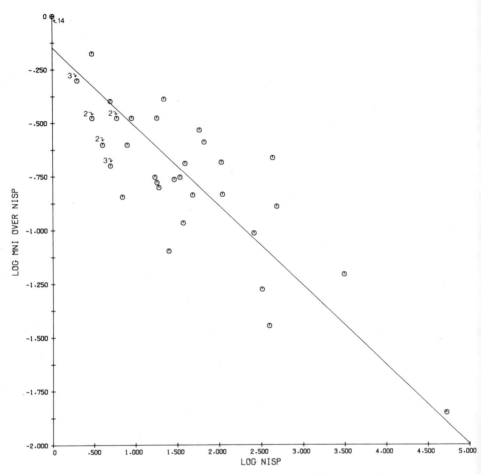

Figure 2.16 The relationship between $\log_{10}$ MNI/NISP and $\log_{10}$ NISP for the Buffalo site birds and mammals.

relationship is MNI/NISP = $0.71(\text{NISP})^{-.37}$ ($r = -0.89$, $p < .001$). Here, the number of individuals defined per specimen decreases at a decreasing rate as the number of identified specimens per taxon increases. The relationship between $\log_{10}$ NISP/MNI and $\log_{10}$ NISP for Prolonged Drift (Gifford *et al.* 1980) is shown in Figure 2.17; the corresponding equation is NISP/MNI = $2.06(\text{NISP})^{.36}$ ($r = 0.84$, $p < .001$). Here, the number of specimens per individual increases at a decreasing rate as the number of specimens per taxon increases. The regression equations for the relationship between MNI/NISP and NISP for the sites listed in Table 2.16 are shown in Table 2.20. I also note that Casteel (1976/1977) examined this relationship in his composite sample of 610 data points

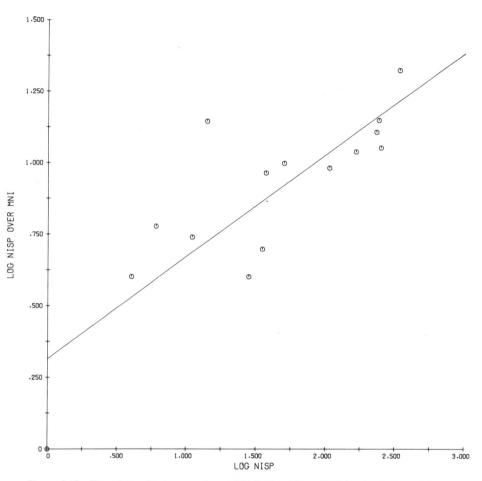

Figure 2.17 The relationship between $\log_{10}$ NISP/MNI and $\log_{10}$ NISP for the Prolonged Drift mammals.

and found the regression line for this sample to be described by MNI/NISP = $0.77(\text{NISP})^{-.48}$. As Casteel noted, his constants were within the range of those I had earlier presented on the basis of the analysis of six separate faunas (Grayson 1974a).

Thus, the relationship between MNI/NISP and NISP is hyperbolic (Figure 2.16); that between NISP/MNI and NISP is parabolic (Figure 2.17). Gejvall (1969) had, in fact, already noted this relationship within archaeological faunas. As part of an analysis of the Greek archaeological site of Lerna, Gejvall "tried to . . . discover how many [bone] fragments of the size normally found in debris are necessary as a foundation for the minimum number of individuals"

TABLE 2.20

Regression Equations and Correlation Coefficients for the Relationship between MNI/NISP and NISP at Prolonged Drift, Apple Creek, Buffalo, Dirty Shame Rockshelter Stratum 2, Dirty Shame Rockshelter Stratum 4, and Fort Ligonier.

Site	Regression equation	r	p
Prolonged Drift	MNI/NISP = 0.49(NISP)$^{-.36}$	−.834	<.001
Apple Creek	MNI/NISP = 0.69(NISP)$^{-.41}$	−.880	<.001
Buffalo	MNI/NISP = 0.71(NISP)$^{-.37}$	−.888	<.001
Dirty Shame Stratum 2	MNI/NISP = 1.03(NISP)$^{-.36}$	−.918	<.001
Dirty Shame Stratum 4	MNI/NISP = 1.04(NISP)$^{-.38}$	−.976	<.001
Fort Ligonier	MNI/NISP = 1.11(NISP)$^{-.60}$	−.888	<.001

(Gejvall 1969:4). Using data for pig, sheep–goat, and cow derived from Lerna and, apparently, several other sites, Gejvall plotted MNI/NISP values expressed as a percentage against NISP. Finding that the resultant relationship was curvilinear, Gejvall fit a curve to his data by inspection. Since that curve seemed to become approximately parallel to the X axis when NISP exceeded 300, he concluded that at least 300 specimens per taxon are needed for valid minimum number analysis. Had Gejvall calculated the equation best describing the relationship between MNI/NISP and NISP at Lerna, he would have realized that MNI/NISP values within his sample continued to decrease as specimen counts increased (at Lerna, MNI/NISP = 0.69(NISP)$^{-.35}$; see Grayson 1978a), though the sample size he chose, 300 specimens, was well beyond the inflection point of the relationship he described. Gejvall (1969) had made a valuable point: MNI/NISP values are a function of sample size.

Shotwell (1958) had investigated this relationship even earlier. Shotwell (1958:276) examined the "percent of difference in the number of specimens per individual" and MNI by plotting corresponding values on log–log paper. From these plots, Shotwell (1958:275) concluded that the "percent of difference in the number of specimens per individual due to a difference of one in the determination of the minimum number is high when the minimum number is less than three; above three, the difference is greatly reduced." That is, Shotwell argued that reduction in NISP/MNI values were greatest for taxa represented by three individuals or less, and concluded that the use of any measure involving NISP/MNI should be confined to larger samples. Shotwell fully realized that the relationship between NISP/MNI and NISP was curvilinear, but felt that the effects of this relationship would be sufficiently dampened for taxa represented by more than three individuals that the measure could be used to quantify relative skeletal completeness in paleontological faunas.

Thus, the refinements introduced by Casteel and myself into the examination

TABLE 2.21
Numbers of Identified Specimens, Minimum Numbers of Individuals, and NISP/MNI Values for Deer and Woodchuck at the Buffalo Site

Taxon	NISP	MNI	Observed NISP/MNI	Predicted NISP/MNI
Deer	52896	746	70.91	78.36
Woodchuck	109	16	6.81	7.95

of the relationship between MNI/NISP and NISP simply represent more precise measurement of things already known as a result of the work conducted by Gejvall in archaeology and Shotwell in paleontology. Nonetheless, that the relationship between these two variables precludes straightforward analysis of "the number of individuals per specimen" or "the number of specimens per individual" remains insufficiently appreciated by both paleontologists and archaeologists (see, for instance, the suggestions in Schiffer 1983; Shipman 1981; and Wing and Brown 1979). In particular, what seems to be insufficiently appreciated is the fact that both MNI/NISP and NISP/MNI measure sample size.

This fact may already be clear, but an examination of the Buffalo and Prolonged Drift faunas (Figures 2.16 and 2.17) will make it clearer still. Let us say that we had hypothesized that the deer and woodchuck represented at the Buffalo site (Table 2.21) had been butchered in different ways, that the larger deer had been partitioned into smaller pieces, and thus more specimens per individual, than the smaller woodchuck. The most obvious way to test this hypothesis is to divide the number of identified specimens by the minimum number of individuals for each taxon, and compare the results. If we do that, we get the results listed in the third column of Table 2.21. Our hypothesis seems strongly confirmed: for every individual of deer, there are 70.91 identified specimens, but for each individual of woodchuck, there are only 6.81 identified specimens.

The flaw in our reasoning, however, is that we have neglected the curvilinear relationship between NISP/MNI and NISP. As NISP increases, values of NISP/MNI increase strictly as a function of sample size. The fact that there are many more identified specimens per individual for deer than for woodchuck may simply represent the fact that there are many more deer bones than woodchuck bones. The question we should be asking is whether there are more deer bones per individual than woodchuck bones per individual given the number of identified specimens for each taxon. That question can be answered by using the equation that relates NISP/MNI to NISP for this site (NISP/MNI = $1.40(\text{NISP})^{.37}$), and determining the number of identified specimens per individual that is expected given this relationship. For deer, the answer is 78.36 specimens per individual; for woodchuck, it is 7.95 specimens per individual (Table 2.21, Column 4). The comparison between observed (Table 2.21. Column 3) and expected (Column 4) values shows that our hypothesis is not well supported at all. There are 17% fewer woodchuck specimens per individual than predicted, and 11% fewer deer specimens per individual than predicted. Thus, taking the curvilinear relationship between NISP/MNI and NISP into account prevents us from confusing a sample size effect with support for our hypothesis.

Inspection of Figure 2.18, which plots $\log_{10}$ NISP/MNI against $\log_{10}$ NISP for

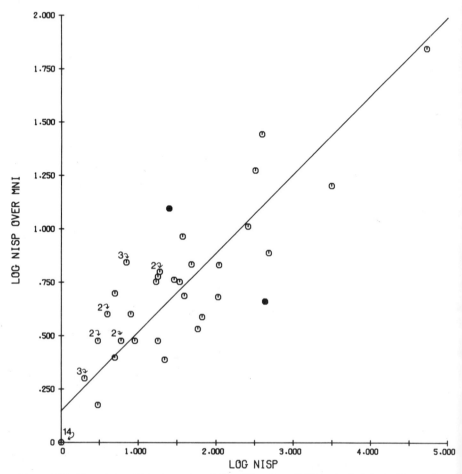

Figure 2.18 The relationship between $\log_{10}$ NISP/MNI and $\log_{10}$ NISP for the Buffalo site birds and mammals; outliers indicated by closed circles.

the Buffalo site, shows that there are some oddities in this assemblage. In particular, there are two taxa whose NISP/MNI values differ from expected values by more than two standard deviations. The NISP/MNI value for raccoon is 4.60, but the predicted value is 13.24. For some reason, there are many fewer specimens per individual of raccoon here than would be expected by chance. Similarly, the NISP/MNI value for elk is 12.50 but the predicted value is 4.61: there are nearly three times as many specimens per individual than one would predict for this fauna. I am unable to extract any clues from Guilday (1971) as to why this might be the case, but at least I am certain that these values are not determined by sample size.

The same kind of analysis can, of course, be applied to Prolonged Drift, but I will simply note that the relationship of MNI/NISP to MNI within this fauna reveals a single outlier (Figure 2.19; in all regression analyses reported here, I am treating as outliers those observed values that deviate from predicted values by two or more standard deviations). Warthog is represented by 14 specimens but by only one individual, or by an MNI/NISP value of 0.071. The predicted MNI/NISP value, however, is 0.186. There are significantly fewer individuals defined per specimen of warthog at Prolonged Drift than one would predict from the fauna as a whole. Although Gifford *et al.* (1980) do not discuss these remains, reserving their detailed analysis for taxa represented by larger samples, their report does list the elements identified for each taxon. The 14 identified warthog specimens are all lower-limb elements and vertebrae, suggesting that butchering practices may well have played a role in producing the very low MNI/NISP value for this animal.

All faunal analysts should be attuned to all uses of NISP/MNI or its reciprocal, since these ratios will primarily reflect sample size if steps are not taken to remove sample size effects. How harmful these effects can be is readily seen by examining J. Arnold Shotwell's method of paleoecological reconstruction.

Shotwell's attempts to infer the community membership of the taxa represented within mammalian paleontological assemblages (Shotwell 1955, 1958, 1963) have become classic, because they were pioneering efforts to extract paleoecological information from paleontological data. They gained much attention in both paleontology and archaeology, and continue to do so now (e.g., Behrensmeyer 1975; Dodson 1971; Estes and Berberian 1970; Holtzman 1979; Shipman 1981; Simpson 1965; Thomas 1971; Wilson 1960). Shotwell's method attempts to group together those mammals represented in paleontological assemblages that lived together in a single community. Shotwell assumed that the skeletons of animals that lived closest to the site of deposition will be more fully represented in the assemblage than the skeletons of animals that lived farther from the site: "the community of the mammals with the greater relative skeletal completeness is the one nearest to the site of deposition" (Shotwell, 1958:272). Shotwell referred to this set of animals as the proximal

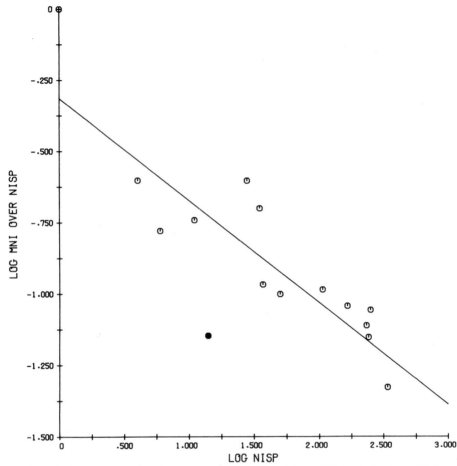

Figure 2.19 The relationship between $\log_{10}$ MNI/NISP and $\log_{10}$ NISP for the Prolonged Drift mammals; outlier indicated by closed circle.

community, as contrasted with communities that existed further from the site of deposition.

The heart of Shotwell's method involves the use of the "corrected number of specimens per individual" as a measure of community membership. This index is, in turn, composed of two separate but interrelated measures: the minimum number of individuals per taxon, and the corrected number of specimens per taxon.

Shotwell (1958:272) defined the minimum number of individuals as "that number of individuals which are necessary to account for all of the skeletal elements (specimens) of a particular species found in the site." He assumed that this measure would provide an indication of the relative density of animals within the proximal community. The corrected number of specimens per taxon was calculated as

$$\frac{\text{NISP}(100)}{\text{Estimated Number of Elements}}$$

in which the estimated number of elements for a given taxon refers to "the number [of elements] less vertebrae and ribs which would be expected if all the elements of the skeleton that can be recognized were recovered" (Shotwell 1958:272). Shotwell thus not only recognized the potential effects of differential identifiability on NISP, but also attempted to remove those effects by norming NISP values with the number of identifiable elements composing the skeletons of members of each taxon. For a series of taxa with the same number of identifiable specimens, then, those with a greater number of identifiable elements would have a lower corrected number of specimens.

The corrected number of specimens per individual (CSI) was defined by Shotwell as the corrected number of specimens divided by the minimum number of individuals. Shotwell used this index as his measure of relative skeletal completeness. Because members of the proximal community are assumed to be more skeletally complete than those derived from more distant communities, the higher the CSI value for any given taxon, the greater the likelihood that members of that taxon belonged to the proximal community. Shotwell delimited the proximal community by assigning to that community those taxa with CSI values above the average CSI for the entire assemblage. The nature of the proximal community so defined could then be more fully probed through, for example, analyses of the minimum numbers and morphology of the members of that community.

From this description of Shotwell's method, it should be clear that his approach is fully dependent on the corrected number of specimens per individual, a measure that is simply NISP/MNI corrected for the differential identifiability of skeletal elements across taxa. Since NISP/MNI increases at a decreasing rate as NISP increases, it can be expected that Shotwell's CSI will behave in a similar fashion and that, as a result, those taxa with high CSI values will be those taxa that are represented by relatively large numbers of identified specimens in any given fauna.

Shotwell (1958) analyzed six mid-Pliocene paleontological assemblages using CSI to discriminate proximal from more distant mammalian communities. The largest of these six faunas were the Boardman, McKay, and Westend Blowout faunas from northeastern Oregon, and the Hemphill fauna from the Texas panhandle. Plotting CSI values against NISP values for these four faunas demonstrates that CSI values vary with sample size, as predicted. The equations describing these relationships are presented in Table 2.22, while Figures 2.20

TABLE 2.22
Regression Equations and Correlation Coefficients for the Relationship between CSI and NISP for the Boardman, Hemphill, McKay, and Westend Blowout Faunas

Site	Regression equation	r	p
Boardman	CSI = 1.77(NISP)$^{.53}$	.885	.001
Hemphill	CSI = 1.89(NISP)$^{.48}$	.853	.001
McKay	CSI = 2.71(NISP)$^{.45}$	.700	.001
Westend Blowout	CSI = 1.43(NISP)$^{.65}$	.870	.001

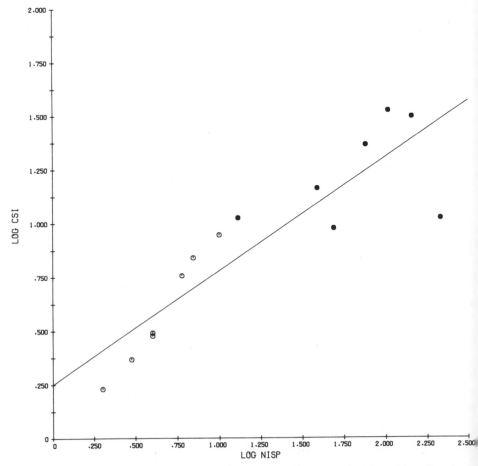

Figure 2.20 The relationship between $\log_{10}$ CSI and $\log_{10}$ NISP for the Boardman fauna. Closed circles represent taxa assigned to the proximal community by Shotwell (1958).

and 2.21 illustrate these relationships for the Boardman and Hemphill faunas. It is clear that CSI and NISP are strongly related, and that the values of CSI are dependent on those of NISP: the higher the sample size per taxon, the higher the CSI value (see Grayson 1978b for illustrations of the relationship of MNI/NISP to NISP for Boardman and Hemphill).

Since higher CSI values define members of the proximal community in Shotwell's method, and since CSI values are a function of NISP, it follows that within Shotwell's four large faunas, taxa assigned to the proximal community (indicated by closed circles in Figures 2.20 and 2.21) are those represented by larger numbers of identified specimens. At Boardman, the members of the proximal

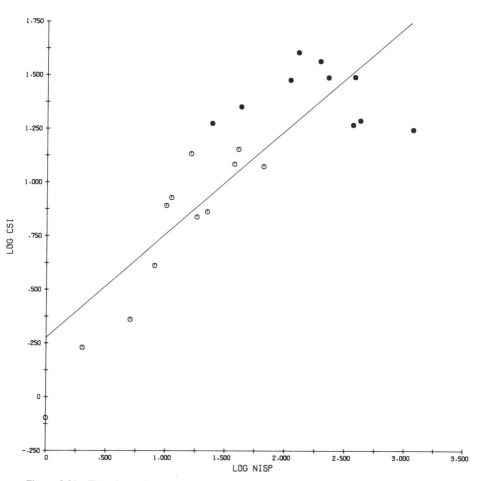

Figure 2.21 The relationship between $\log_{10}$ CSI and $\log_{10}$ NISP for the Hemphill fauna. Closed circles represent taxa assigned to the proximal community by Shotwell (1958).

TABLE 2.23

Average Number of Identified Specimens per Taxon within Shotwell's Proximal and More Distant Communities: Boardman, Hemphill, McKay, and Westend Blowout

Site	$\bar{X}$ NISP: Proximal taxa	Distant taxa
Boardman	91.43	5.14
Hemphill	302.40	19.58
McKay	65.22	9.83
Westend Blowout	68.71	7.56

community average 91.43 specimens per taxon, while the remaining taxa are represented by an average of 5.14 specimens per taxon. The comparable figures for all of Shotwell's large quarry sites are provided in Table 2.23 (these figures differ from those in Grayson 1978b and follow Shotwell 1958).

Thus, Shotwell's measure of community membership is strongly dependent on the numbers of identified specimens per taxon. Only if community membership may be inferred from simple counts of identified specimens per taxon in a may be inferred from simple counts of identified specimens per taxon in a paleontological collection can his method of paleocological reconstruction distinguish members of proximal from more distant communities. If CSI values are to be the focus of analysis, then, as I discussed for NISP/MNI values, the residuals must become the target of study. At Hemphill, for instance, *Pliohippus* has a CSI value that falls more than two standard deviations from the predicted CSI value (an observed CSI of 17.8, as opposed to an expected CSI of 57.48; I note, however, that lower $\log_{10}$ CSI values are overpredicted by my equations, perhaps as a result of Shotwell's corrections for identifiability, and that a higher order equation would provide a better fit for these data). In general, given a plot of $\log_{10}$ CSI against $\log_{10}$ NISP, those points that fall above the best-fit line describing the relationship of these variables to one another represent taxa whose relative skeletal completeness (bones per individual) is greater than those that fall beneath the line, while the line itself represents predicted skeletal completeness across all sample sizes.

Shotwell's important attempts to segregate proximal from more distant communities thus fail because the value NISP/MNI formed the heart of those attempts, and the effects of sample size on that ratio were not taken into account. David H. Thomas's insightful attempt to use Shotwell's approach to distinguish natural from cultural bones in archaeological sites fails for the same reason.

In order to use Shotwell's method to distinguish taxa deposited by human activities in an archaeological sites from those present as a result of noncultural events, Thomas (1971) inverted Shotwell's operating assumption. Thomas cor-

THE RELATIONSHIP OF MNI/NISP TO NISP

rectly noted that, in Shotwell's framework, CSI indicates only the degree of skeletal disruption for each species, but does not in itself assign any causes for that disruption. Thomas (1971:367) argued that, because "the dietary practices of man tend to destroy and disperse the bones of his prey-species," those taxa represented by more complete skeletons (more specimens per individual) most likely represent those that are present as a result of natural causes. Thomas thus focused on low CSI values as indicative of human activities, as opposed to Shotwell, who focused on high CSI values as indicative of membership in a proximal faunal community.

In order to emphasize the disturbed end of a set of CSI values for any given fauna, Thomas defined a coefficient B as $B = 5 - \log_e$ CSI, and used this coefficient as his measure of relative skeletal completeness. He defined CSI precisely as Shotwell (1958) defined it. Thomas then determined the values of B for the mammalian taxa from three Great Basin rockshelters, all located in northwestern Nevada: Hanging Rock Shelter, Little Smoky Creek Shelter, and Smoky Creek Cave. He used the B value for *Lepus* and *Sylvilagus* as his cutoff point to demarcate natural from cultural bone, a decision based on the well-known importance of hares and rabbits as food among Great Basin native peoples during early historic times. Thomas then attributed those taxa with B values equal to or greater than the lowest lagomorph B value to human activities, and those with B values less than the lowest lagomorph B value to nonhuman, "natural" causes.

By adopting Shotwell's method, of course, Thomas also adopted the problems associated with CSI. Figure 2.22 illustrates the relationship between Thomas's B and NISP for the Hanging Rock Shelter mammalian fauna. Here, the relationship between B and NISP is described by $B = 4.45(\text{NISP})^{-.63}$ ($r = -0.92$, $p < .001$), and Thomas assigned to the "proximal" community (that is, to taxa present for natural reasons) those taxa with high CSI, and therefore low B, values. The same pattern is shown by Little Smoky Creek Shelter and by Smoky Creek Cave; the regression equations and correlation coefficients for the relationship between B and NISP are provided in Table 2.24. Were B values to be used to distinguish natural from cultural bones in archaeological sites, the analysis would have to focus on the residuals, defined from the general relationship between B and NISP within each fauna.

Thus, NISP/MNI and any measure derived from this ratio is dependent on the number of identified specimens per taxon, unless the effects of NISP values are removed. In the examples presented here, from 70% (Prolonged Drift) to 81% (Connley Cave No. 4 $\text{MNI}_{10\text{cm}}$) of the variance in NISP/MNI and in MNI/NISP is explained by the number of specimens per taxon. Again, these results are not surprising, since NISP is entered on both sides of the equation. While this ratio may provide valuable information in some settings, it cannot do so unless the effects of NISP are removed. One way to do this is to focus attention on the

TABLE 2.24

Regression Equations and Correlation Coefficients for the Relationship between Thomas' Coefficient B and NISP at Hanging Rock Shelter, Little Smoky Creek Shelter, and Smoky Creek Cave

Site	Regression equation	r	p
Hanging Rock Shelter	$B = 4.45(NISP)^{-.63}$	$-.921$	$<.001$
Little Smoky Creek Shelter	$B = 4.27(NISP)^{-.52}$	$-.868$	$<.001$
Smoky Creek Cave	$B = 4.36(NISP)^{-.54}$	$-.922$	$<.001$

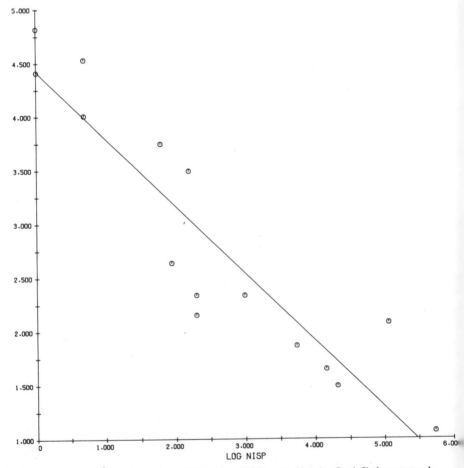

Figure 2.22 The relationship between B and $\log_e$ NISP for the Hanging Rock Shelter mammals.

TABLE 2.25

Regression Equations and Correlation Coefficients for the Relationship between MNI/NISP and NISP for the Connley Cave No. 4 Mammals

MNI	Regression equation	r	p
10 cm	MNI/NISP = $1.10(NISP)^{-.29}$	−.900	<.001
Stratum	MNI/NISP = $1.04(NISP)^{-.46}$	−.893	<.001
Site	MNI/NISP = $0.80(NISP)^{-.47}$	−.848	<.001

scatter of points about the line that best describes the relationship between NISP/MNI (or MNI/NISP) and NISP for the fauna at hand, since that line represents predicted NISP/MNI (or MNI/NISP) values for any given sample size. The residuals—the variance in NISP/MNI or MNI/NISP unexplained by NISP—then become the target of analysis.

Although the analysis of residuals can take into account the effects of NISP on NISP/MNI ratios, it cannot take into account the effects of aggregation on minimum numbers, and it should be clear that aggregation will affect the value of those residuals. Different methods of aggregation affect the slope of the relationship of MNI/NISP to NISP, with slope coefficients decreasing as approaches to aggregation become more agglomerative (see the illustrations in Grayson 1978a). This in itself might not affect the values of the residuals were it not for the fact that minimum number values are differentially altered across taxa as approaches to aggregation (and as slopes) change. As a result, MNI/NISP and NISP/MNI values are differentially altered, and the magnitude of the residuals changes.

One example will suffice. The relationship of MNI/NISP to NISP for the Connley Cave No. 4 mammals is shown in Figures 2.23 (MNI_{10cm}), 2.24 ($MNI_{stratum}$), and 2.25 (MNI_{site}); the corresponding regression equations and correlation coefficients are presented in Table 2.25. When the Connley Cave No. 4 mammals are aggregated by 10-cm units, there are no MNI/NISP residuals at or above two standard deviations. When these mammals are aggregated according to stratum, a single taxon emerges as an outlier. The predicted $MNI_{stratum}$/NISP value for bison is 0.23; the observed value is 0.07, more than two standard deviations beneath the predicted value. When the entire Connley Cave No. 4 mammalian fauna is treated as a single aggregate, there is still a single outlier, but it is not bison. Instead, it is deer, with a predicted MNI_{site}/NISP value of 0.25, and an observed value of 0.08, again more than two standard deviations beneath the predicted value.

Clearly, the approach to aggregation will affect the value of MNI/NISP residuals. While the effects of NISP on NISP/MNI (or MNI/NISP) can be taken into

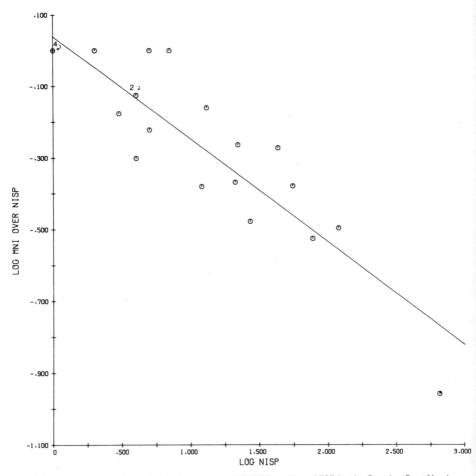

Figure 2.23 The relationship between $\log_{10}$ MNI/NISP and $\log_{10}$ NISP for the Connley Cave No. 4 mammals: MNI_{10cm}.

account through the analysis of residuals, the effects of aggregation cannot. If this ratio is going to be used for any purpose, then the effects of NISP on its value must be controlled through regression. In this way, at least one problem will have been avoided: the analyst will no longer run the risk of confusing those differences in the number of specimens per individual that are due to differing numbers of identified specimens per taxon with those that are due to other causes. But the effects of aggregation will still be uncontrolled, and the possibility that significant NISP/MNI (or MNI/NISP) residuals have resulted at least in part from the selection of aggregation units must be seriously considered.

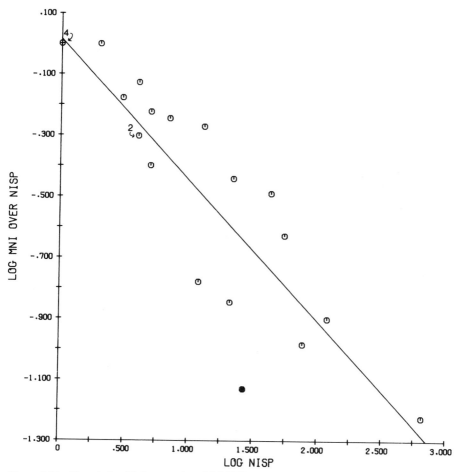

Figure 2.24 The relationship between $\log_{10}$ MNI/NISP and $\log_{10}$ NISP for the Connley Cave No. 4 mammals: $MNI_{stratum}$; outlier indicated by closed circle.

Some Other Approaches to Counting

During the last few years, several new approaches to quantifying taxonomic abundances have been suggested. These approaches reject both minimum numbers and specimen counts as basic units of quantification. Because these approaches are new, they have yet to be applied to a wide variety of vertebrate faunas, a situation that stands in contrast to minimum numbers and specimen counts, both of which developed as intuitively satisfying measures of abundance as vertebrate paleontology itself developed during the nineteenth cen-

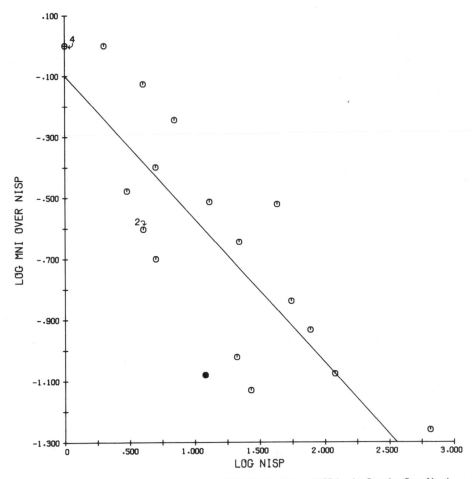

Figure 2.25 The relationship between $\log_{10}$ MNI/NISP and $\log_{10}$ NISP for the Connley Cave No. 4 mammals: MNI_{site}; outlier indicated by closed circle.

tury. Although neither approach is without very serious difficulties, the attempt to construct new methods is of great importance, and I briefly discuss these approaches here.

Matched-Pairs Methods

In 1968, Grover S. Krantz published a short paper in which he expressed dissatisfaction with both specimen counts and minimum numbers as measures of taxonomic abundance. He found specimen counts flawed because, while "comparison of . . . total bone counts should indicate the *relative* numbers

SOME OTHER APPROACHES TO COUNTING

of each game animal taken by the hunters at the time, [it] will give no indication of the *absolute* numbers represented" (Krantz 1968:286; emphasis in original). He also observed that specimen counts were prone to bias because of the possibility of the differential recovery of bones belonging to different taxa. He objected to minimum numbers because there is no way of knowing how many of the bones originally deposited in a site are subsequently available to the faunal analyst to be counted: "the bone counter may have 50% of the original number of bones of one species and 60% of the bones of another without knowing of this difference in preservation" (1968:287).

In the place of specimen counts and minimum numbers, Krantz suggested that the analyst focus on elements that are paired in the skeleton of living mammals, such as the mandible, and that the *absolute* number of animals N that had contributed to a faunal assemblage be estimated as

$$N = \frac{R^2 + L^2}{2P}$$

in which R represents the total number of right specimens in the assemblage for the element involved, L the number of left specimens, and P the number of perfectly matched pairs that exist among the rights and lefts.

Casteel (1977) later criticized this approach as inferior to traditional minimum number calculations. As Fieller and Turner (1982) and Wild and Nichol (1983) have pointed out, however, traditional minimum numbers and Krantz' matched-pairs method have very different goals. While minimum numbers simply attempt to summarize the number of animals that are needed to account for a given assemblage of bones, Krantz' method was meant to estimate the number of animals that had *originally* contributed to the collection.

Recent examinations of Krantz' approach have suggested that modifications be made to the formula he presented. These modifications derive from work done in ecology, where measures of this sort are in common use in capture–recapture studies. Fieller and Turner (1982) recommended that

$$N = \frac{RL}{P}$$

be used, while Wild and Nichol (1983) suggested that

$$N = \frac{(R+1)(L+1)}{P+1} - 1$$

might be more appropriate for small samples.

Although Fieller and Turner (1982) are very optimistic about the potential of this approach for estimating the number of animals that contributed to a faunal assemblage, Wild and Nichol (1983) are more pessimistic. Pessimism is fully appropriate. As Wild and Nichol (1983) observe, the method demands that the analyst be able to discover all matched pairs for any given element that remains in the site at the time of excavation. This demand poses two problems. First, since an unmatched bone whose partner has simply not been collected has very different meaning from an unmatched bone whose partner has disappeared since being deposited, the approach requires that all bones in a site be retrieved. Second, the method poses the currently unsolved technical problem of discovering all matched pairs in the faunal collection that has been retrieved. It is no accident that Krantz (1968) chose mandibles to introduce the approach, no accident that Wild and Nichol (1983) used fish dentaries and premaxillae to illustrate the approach, and no accident that Fieller and Turner (1982) used modern otoliths that came from known individuals for their illustration and chose mandibles and skulls in discussing pair-matching in general. Wild and Nichol (1983) noted they even had difficulties in matching dentaries and premaxillae, and archaeologists, who generally work with fragmented material, have no means of validly extracting true matched pairs for postcranial material (see, for instance, the discussion of bilateral asymmetry in Jolicoeur 1963 and the references in Goss 1983). In the ecological setting, mice and birds are tagged to allow recaptured individuals to be recognized, but the faunal setting does not provide this convenience. As both Fieller and Turner (1982) and Wild and Nichol (1983) note, falsely matched and falsely unmatched pairs result in inaccuracies in estimation using this method. Although difficulties of this sort might be remedied by changes in excavation procedures and by technical advances in recognizing which bones came from which individuals, the method also requires that matched pairs were initially deposited in the site, and that any unmatched specimens are unmatched because their partners have decayed. Such an assumption flies in the face of all that we have learned from taphonomic studies during the past decade (see, for instance, the discussion in Binford 1978, 1981). To the extent that this assumption has been violated for any given faunal assemblage, the results are inaccurate.

The main advantage of Krantz' approach, as modified by Fieller and Turner (1982) and Wild and Nichol (1983), would seem to be provided by the fact that, unlike either specimen counts or minimum numbers of individuals, the statistical basis of the approach has been well explored. Nonetheless, the problems posed by matched-pairs methods are formidable, and I conclude with Wild and Nichol (1983:344) that "in practice its usefulness is severely limited."

Minimal Animal Units

To this point, the minimum numbers I have discussed depend on the definition of a most abundant element. The minimum number of individuals is calculated as the smallest number of individual animals needed to account for that

SOME OTHER APPROACHES TO COUNTING

most abundant element. When these numbers are further manipulated, the analyst is forced to assume that the entire individual forms the appropriate unit of analysis.

Binford (1978) has expressed reservations about this assumption, noting, for instance, that meat is not utilized by people in units of single animals, but instead in units of animal segments (see also Lyman 1979). He correctly observes that minimum numbers obscure the existence of such segmental units, and rejects the use of minimum numbers in subsistence analyses as a result. In its place, he substitutes a very different measure, which he also calls the minimum number of individuals: *"all MNIs will be calculated by dividing the observed bone count for a given identification unit by the number of bones in the anatomy of a complete animal for that unit"* (Binford 1978:70; emphasis in original). Thus, for a collection of 70 right and 30 left femora of a given mammalian taxon, Binford would calculate an "MNI" of 50; that is, 100 femora divided by the number of femora that occur in the mammalian skeleton. This value is, of course, quite different from the standard MNI value of 70 that would be calculated for this collection (if age and size are not taken into account).

It is surprising that Binford initially chose to call his values "minimum numbers of individuals," since they have nothing to do with numbers of individuals. More appropriate is the label "minimal animal units" (MAU) which he has recently applied (Binford 1984). It should be clear that rather than being minimum numbers of individuals, MAUs are specimen counts normed by the number of times the element involved is represented in the skeleton of the taxon involved. In some cases, such as those in which virtually entire skeletons are preserved, Binford's MAUs and standard MNIs may be virtually identical (e.g., Thomas and Mayer 1983), but Binford's approach is meant to provide "undistorted conversions of the actual count of bones into animal units' (1978:70).

Because Binford's MAUs are simply normed specimen counts, they may avoid many of the problems that affect MNIs. These counts do, however, possess some problematic aspects. First, it seems clear that if an analyst wants to count segmental units, all the analyst really needs to do is to count specimens. How many femoral segments are present in the assemblage of 70 right and 30 left femora noted above? It is clearly not 70 (the minimum number count), but it is also not 50 (Binford's count). It is 100. Bones such as phalanges and vertebrae, however, present a more perplexing problem. How many segmental units are represented by 80 first phalanges from an animal whose skeleton has eight of these elements? What constitutes the appropriate piece of anatomy to use in deriving MAUs here? Is it the single phalanx? If so, there are 80 MAUs here. Are the first phalanges of the fore- and hindlimbs to be treated separately? If so, then there are 20 MAUs here. Are all the first phalanges to be considered together? If so, then there are 10 MAUs in our collection. Binford uses the last of these figures, but it should be clear that his conversion requires an assumption of only slightly less magnitude than assuming that an entire individual was deposited in the site. If the Nunamiut behave in such a way as to allow this assumption, did whatever accumulated the bones at Olduvai Gorge (Binford 1981) or Klasies River Mouth Cave No. 1 (Binford 1984) behave in this way as well? Binford's approach may obscure segmental units less than do minimum numbers, but obscure them it must.

There is an even greater problem here. Binford's approach often requires that an additional analytic step be taken. Archaeologists rarely have complete bones or complete proximal and distal ends of bones with which to work. Instead, they have fragments of proximal and distal femora, fragments of skulls, fragments of mandibles. If we had, say, 100 badly fragmented proximal femora in an assemblage, how many minimal animal units are present? If we simply divide by two, we ignore the fact that many of our fragments may have come from the very same bone. How do we handle this problem? Binford handles it by reconstructing "minimum numbers of elements" (MNE). He apparently took this step for the Combe Grenal fauna (Binford 1981: Table 4.03), and makes heavy use of such reconstructions in his analysis of the Klasies River Mouth Cave No. 1 fauna (Binford 1984; for insightful discussions of this fauna, see Klein 1975, 1976, 1977, 1978, 1980, 1981, 1982, 1983). It should be clear that once this step has been taken, Binford's minimal animal units are as fully prone to aggregation effects as are minimum numbers of individuals, since the first step in calculating minimum numbers of elements must be the determination of a set of appropriate analytic clusters. Klasies River Mouth Cave No. 1 provides a perfect example. Binford (1984) chose to treat this entire faunal sample, which accumulated across tens of thousands of years, as a single faunal aggregate! Had he aggregated his materials in a different fashion — by using, for instance, the subdivisions employed by Klein (1976) — there can be no doubt that his results would have been very different.

In short, Binford's minimal animal units represent an interesting, but flawed, attempt to circumvent the problems posed by other measures of taxonomic abundance.

Minimum Numbers and Specimen Counts: Some Conclusions

In the settings I have discussed, minimum numbers of individuals seem to provide an extremely poor measure of taxonomic composition. The effects of aggregation are such that the analyst simply cannot know whether the frequencies of taxa within and among faunas are reflecting taxonomic composition or analytic choices made in the field or in the lab. At Connley Cave No. 4, are there too few bones per individual of deer, of bison, or of neither? At Hidden Cave, have 78% or 48% of the distal tibiae survived? Do the slopes provided by Hesse's data for caprines from Near Eastern tells differ because of aggregation or for some other reason? In all of these cases, the effects of aggregation on minimum numbers are such as to allow us little faith in the meaning of those numbers.

The cases I have discussed are those in which there are choices to be made in the definition of faunal aggregates. In such caves and rockshelters as the Connley Caves and Hidden Cave, the choice is between arbitrary units within strata, strata as a whole, and combinations of strata. In such open sites as Buffalo and Apple Creek, the choice may be of the same sort, but may also include whether or not to aggregate by storage pits, midden units, and house floors. Wherever there are obvious choices of this sort to be made, it is clear that the potential effects of aggregation greatly diminish the validity of minimum numbers as the sole measure of taxonomic abundance. The analyst can no longer be sure that what is being measured is what he or she set out to measure.

But what about situations in which there are no clear choices? The Gatecliff Shelter strata, for instance, are remarkably well-defined and were excavated with exquisite care. Could one not validly use minimum numbers for, say, the Stratum 9 faunal assemblage, which contained 636 specimens of 30 mammalian taxa dating to about 3300 B.P.? Certainly there is no choice as to the aggregate to be used here, since the only possible aggregate is provided by the stratum itself. If minimum numbers are calculated, they must be calculated on a stratum-wide basis. Suppose, however, that this fauna had been collected with even more care than it was. What if Gatecliff had been excavated 50 years later when, we may hope, techiques for defining separate depositional events will be much better defined than they were when Gatecliff was actually excavated, and such an excavation was able to both demonstrate and take advantage of fine stratigraphic subdivisions within Stratum 9? Then, the choice would exist, and aggregation effects would come into play. Minimum numbers, and the relative abundances they indicate, would change, while numbers of identified specimens would remain unchanged. It is not idle to speculate that this might occur. As Thomas and Mayer (1983:355) have noted,

> most of the Gatecliff "living floors" were actually palimpsest accumulations, clearly deposited as multiple events, often separated by minute silt bands between individual occupations . . . we were sometimes unable to follow these microstratigraphic breaks throughout the entire surface, and we were forced to analytically lump these mini-occupations into a single horizon.

The fact that there is no choice to be made as to how aggregate the Stratum 9 fauna does not mean that aggregation effects are absent; it just means that they cannot be detected. In this sense, Stratum 9 at Gatecliff Shelter represents a minute Swartkrans, containing many discrete, but analytically undivided events.

Thus, I argue that the effects and potential effects of aggregation on minimum numbers are such as to make these numbers an extremely poor choice as the basic measure of relative taxonomic abundance. While they may be useful as part of a wider-ranging analysis that is based on the number of identified specimens per taxon, minimum numbers seem flawed at their heart. Specimen counts provide the same sort of information on relative abundances that is provided by minimum numbers, yet are not affected by aggregation. As a result, the number of identified specimens per taxon provides the best unit we have available for measuring the relative abundances of vertebrate taxa in archaeological and paleontological sites.

CHAPTER 3

Levels of Measurement

To this point, I have assumed that the analyst's initial goal in quantifying the taxonomic composition of a vertebrate fauna is the derivation of a ratio level measure of taxonomic composition. Thus my critique of minimum numbers has stemmed from the fact that the effects of aggregation on taxonomic abundances calculated from minimum numbers are such as to differentially alter abundance ratios between and among taxa. Might it not, however, be too much to assume that either minimum numbers of individuals or numbers of identified specimens necessarily provide ratio scale measures of taxonomic abundance within single faunal assemblages?

In both interval and ratio scales, of course, distances that are numerically equal in terms of the measurements themselves must stand for distances that are equal in terms of the variable being measured. These scales are distinguished by the fact that the zero point of an interval scale is arbitrary, while ratio scales originate at a true zero point. Something that has a temperature of 0°C, for instance, still possesses the molecular movement that causes heat, but something that weighs 0 kg doesn't weigh anything at all. In the faunal setting, not only do we search for a measure in which distances numerically equal on the measurement scale stand for distances actually equal in terms of the variables (taxonomic abundances) being measured, but we also seek a measure in which a count of zero means that the taxon was, in fact, absent from the assemblage. What we seek, in short, is a ratio scale measure of taxonomic abundance. With such a scale, we could say that Taxon A, with an abundance measured as 50 units, was twice as abundant as Taxon B, with an abundance measured as 25 units, and two-thirds as common as Taxon C, with an abundance measured as 75 units. Here, I argue that MNI- and NISP-based assessments of the taxonomic composition of single faunal assemblages may provide no more than an ordinal scale measure of that composition. With such an ordinal scale measure, the analyst can specify with certainty only that a given taxon is more or less abundant than another within the assemblage. Indeed, I argue that minimum numbers and specimen counts may even fail here.

MNI and NISP as Ratio Scale Measures

There are a number of ways in which it can be shown that minimum numbers do not provide a ratio scale measure of taxonomic abundance. One depends on the effects of aggregation. When one ratio scale is converted into an equivalent ratio scale based on different units of measurement, the values of the ratios of measurements to one another do not change because both scales have true zero points and because the unit of measurement remains constant in each. The ratios of the weights of two objects, for instance, remain the same if they are measured first in kilograms and then in pounds. I have already discussed the effects of aggregation on the ratios of taxonomic abundance as measured by minimum numbers: those ratios are altered as approaches to aggregation change. Table 2.8 presented such changes for a series of taxa from Connley Cave No. 4; other examples can be found by examining Table 2.10, which presents minimum numbers derived from two different aggregation approaches for the Hidden Cave mammals. Such alterations themselves demonstrate that minimum numbers calculated using different approaches to aggregation do not provide identical ratio scale measures of taxonomic abundance. Because MNI-based measurements possess true zero points, the altered ratios of taxonomic abundance that result from different approaches to aggregation must result from the fact that the unit of measurement itself has not remained constant. One approach to aggregation in a given setting may, of course, provide a valid ratio scale abundance measure but, if so, there is no way of knowing which one it might be.

That minimum numbers are *minimum* numbers provides a second reason for denying them ratio scale status. "Minimum numbers of individuals" means exactly what it says: the minimum number of individual animals needed to account for the faunal material in some aggregation unit. The actual abundances that any given minimum number may represent will vary from that minimum number to some unknown higher figure, the upper limit being determined by the number of identified specimens for that taxon. As a result, the relationship between the distance between the minimum numbers for two taxa in a faunal assemblage and the actual abundances of those taxa is unknown. It is not possible, for instance, to say that an MNI of 40 reflects twice as many individuals as an MNI of 20, or that equal minimum number values for different taxa reflect equal actual abundances. This obvious fact is illustrated in Figure 3.1, which presents minimum numbers, numbers of identified specimens, and "actual" abundances in a hypothetical fauna. When measured by minimum numbers, the ratio of the abundances of Taxon 1 to Taxon 2 in this sample is 0.50, but the ratio of the actual abundances — the parameter we are trying to estimate — is 0.40 (see Table 3.1). Were this fauna reaggregated, some other ratio would, of course, result, the magnitude of the difference between the

MNI AND NISP AS RATIO SCALE MEASURES

TABLE 3.1

Ratios of Taxonomic Abundance for the Taxa Illustrated in Figure 3.1

Taxa	Ratios		
	MNI	NISP	Actual
T_1-T_2	0.50	0.71	0.40
T_1-T_3	0.33	0.56	0.25
T_2-T_3	0.66	0.78	0.63

ratios depending on the nature of the distribution of most abundant elements within the site. Because minimum numbers bear an unknown relationship to actual abundances, and because there is no reason to think that the ratios between minimum numbers and actual abundances remain stable across all taxa within a fauna, it cannot be shown that minimum numbers provide a ratio scale measure of relative taxonomic abundance, and there is every reason to think that they do not provide such a measure.

Clearly, we are not on safe ground in treating minimum numbers as ratio scale units. Can numbers of identified specimens per taxon be treated as

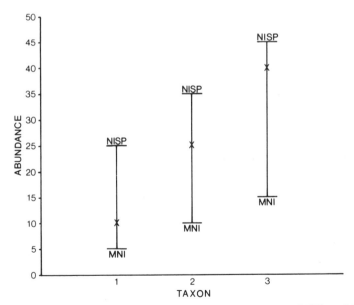

Figure 3.1 Minimum numbers (MNI), numbers of identified specimens (NISP), and "actual" abundances (X) of three taxa in a hypothetical fauna.

providing a ratio scale measure? It is probably already clear that they cannot. While aggregation does not affect specimen counts (unless Binford's approach is used), the relationship between the ratios provided by those counts and actual abundances are as unknown for specimen counts as they are for minimum numbers (see Table 3.1). On the one hand, minimum numbers represent *minimum* numbers and hence do not provide a ratio scale measure of taxonomic abundance. On the other hand, numbers of identified specimens represent *maximum* numbers of individuals, and hence also do not provide a ratio scale measure with any certainty. In either case, if we happen to have obtained a ratio scale measure of abundance, there is no way of knowing it; in both cases, there are many reasons to doubt that such a scale has been obtained.

Both minimum numbers and specimen counts are statistics whose values are meant to provide estimates of "actual" abundances; these estimates are in turn used to measure relative abundances. In many statistical settings, the nature of the distribution of the variables under study is known, and well-based assumptions can be made that allow the accuracy of an estimate to be assessed. There are, for instance, a number of osteological variables—the length of a given bone, say—that are normally distributed. Knowledge of this fact is of great value in attempting to estimate the population values of such a variable from a sample of that population. Minimum numbers and specimen counts, however, attempt to estimate the parameter "abundance" by providing the upper and lower limits, respectively, to the possible distribution of abundance values, while we remain ignorant of the nature of the distribution between these limits. This procedure for estimating taxonomic abundance has little in common with normal statistical practices. Given that we know nothing of the nature of the frequency distribution that begins at MNI and ends at NISP for a set of taxa, this procedure is also not one that can provide us with robust (ratio scale) estimates of taxonomic abundance with any certainty.

MNI and NISP as Ordinal Scale Measures

If on a single-site basis we cannot show that either minimum numbers or specimen counts have provided a ratio scale measure of taxonomic abundance, and if simple considerations of the nature of faunal data suggest that they are not likely to provide us with such a measure, or at least that we have no way of knowing it if they do, can we feel at all secure that either MNI or NISP provides us with an ordinal measure? Is there any way of knowing that either measure is even providing accurate information on the rank order of taxonomic abundance within a given faunal assemblage, without worrying about how much more abundant one taxon is than another? If minimum numbers fail on this score as well, all we could really feel secure about is that the taxa represented on a taxonomic list were, in fact, represented in the fauna from that site.

MNI AND NISP AS ORDINAL SCALE MEASURES

It is not difficult to show that both minimum numbers and specimen counts may provide valid ordinal level measures of taxonomic abundance for a given faunal assemblage. More precisely, it is not difficult to show that one can get identical, or extremely similar, sets of rank orders for a faunal assemblage when both specimen counts and minimum numbers are used to derive those rank orders. The reason for the frequency with which this occurs is fairly simple. The distribution of taxonomic abundances within vertebrate faunas is routinely one in which very few taxa are represented by a large number of specimens or individuals, while most are represented by small numbers of these units. The nature of the distribution of taxonomic abundances for the faunas listed in Table 2.16 is illustrated in Figures 3.2 through 3.7; additional examples can be seen in Grayson (1979b). In each of these cases, as taxonomic abundances increase, the magnitude of differences in abundance between adjacent taxa also increases, whether measured by MNI or NISP.

There are a number of reasons why such a distribution might characterize so many archaeological faunas. Those relatively few animals that are present in great abundance may often represent economically important taxa (cow, pig, and sheep at Fort Ligonier; raccoon and deer at Apple Creek). The relatively numerous taxa that are poorly represented may reflect economically unimportant taxa, animals whose presence is adventitious (for instance, five specimens of ground squirrel, chipmunk, and rice rat at Apple Creek), or animals whose

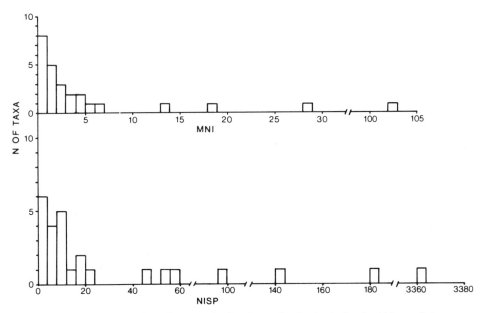

Figure 3.2 The distribution of taxonomic abundances for the Apple Creek midden and plow-zone mammals.

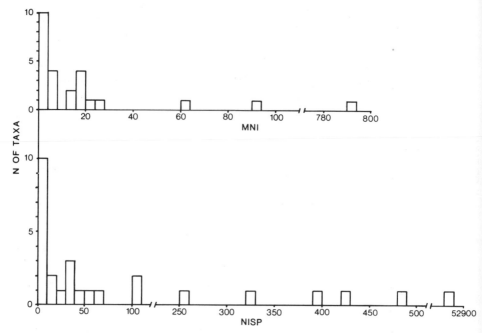

Figure 3.3 The distribution of taxonomic abundances for the Buffalo site mammals.

abundance has been artificially reduced by collecting techniques (e.g., Thomas 1969).

No matter what the reason, however, such a distribution does serve to increase the probability that the ordinal abundances of common taxa may be accurately reflected by either minimum numbers or specimen counts. As regards minimum numbers, taxa that are widely separated in abundance are less likely to have their rank orders rearranged by different approaches to aggregation. At the lower end of the distribution, changes in absolute abundance due to aggregation may reverse rank orders, but they are more likely to cause increases or decreases in ties. Although increases or decreases in ties will most certainly affect ordinal statistical techniques, it is also true that taxa at the lower end of the distribution are so poorly represented that it is questionable whether they should be treated in anything other than a nominal, presence/absence sense. Indeed, such taxa are often used only to create larger classes of taxa based on nonphylogenetic criteria that are then manipulated as units — "mesic" versus "xeric" taxa, for instance (e.g., Grayson 1977b; Harper and Alder 1970). Once this is done, those taxa represented by relatively small numbers of specimens or individuals become part of more abundant, taxonomically composite classes that fall toward the upper end of the distribution of

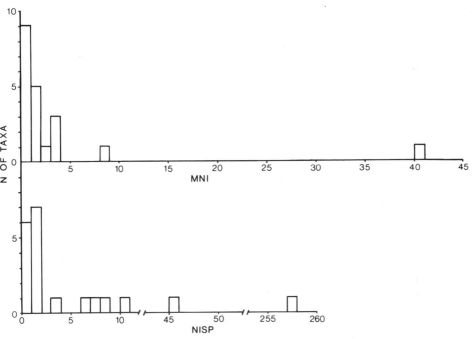

Figure 3.4 The distribution of taxonomic abundances for the Dirty Shame Rockshelter Stratum 2 mammals.

taxonomic abundance. As a result, their rank orders are less prone to being altered as a result of different methods of aggregation.

Similar considerations apply to specimen counts. The greater the differences in specimen-based abundances between taxa, the less the chances that specimen interdependence will alter the rank order abundances of those taxa; the less those differences, the greater the chances. Again, the rank order abundances of well-separated taxa toward the upper end of the distribution may be accurately reflected by specimen counts, while those counts will probably poorly reflect rank order abundances at the lower end of the distribution.

Thus, an argument can be made that both specimen counts and minimum numbers may provide acceptable estimates of rank order abundances of the more common taxa in at least some situations. Rank orders may remain stable between minimum number and specimen counts, and among minimum number counts calculated by different aggregation approaches, even though absolute abundances indicated by these counts may vary widely. The degree of rank order stability will be closely related to the degree of separation of the taxa involved in terms of their MNI- or NISP-based sample sizes.

The Connley Cave No. 4 mammals provide an example. The number of

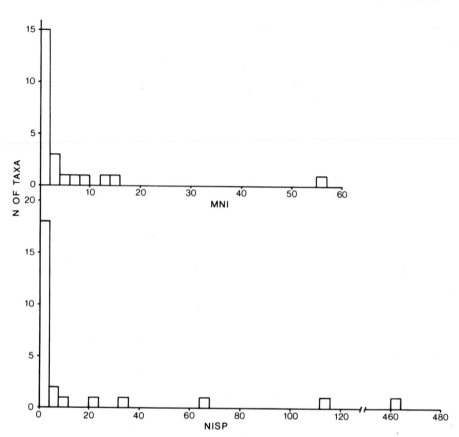

Figure 3.5 The distribution of taxonomic abundances for the Dirty Shame Rockshelter Stratum 4 mammals.

identified specimens per taxon for these mammals is provided in Table 2.4; MNI_{10cm}, $MNI_{stratum}$, and MNI_{site} are provided in Table 2.5. The rank orders of abundance provided by these four measures are given in Table 3.2. It is clear that rank orders have not remained stable across measuring units: both inspection of rank orders and Kendall's tau values show that some ordinal rearrangement has occurred (I have used Kendall's tau here because of the relatively large number of tied ranks; see Kendall 1970). Not surprisingly, the rearrangements are least between those sets of measurements derived from the most similar measurement units, and greatest between the least similar. Thus, the approaches that divide the collection into the largest number of faunal aggregates (NISP and MNI_{10cm}) show relatively minor ordinal rearrangements between themselves, and the same is true for those approaches that divide the

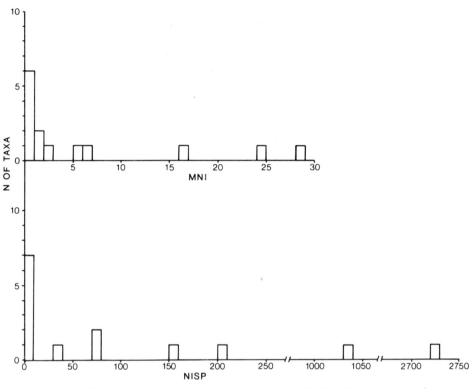

Figure 3.6 The distribution of taxonomic abundances for the Fort Ligonier mammals.

collection into the smallest number of aggregates ($MNI_{stratum}$ and MNI_{site}). The magnitude of rearrangement increases as increasingly agglomerative approaches are compared with increasingly divisive ones. Nonetheless, all Kendall's tau values are associated with probabilities of $<.001$. In no case do the Kendall's tau values exceed those that might occur from resampling the same population using the same aggregation method; the tau values indicate that the same population of rank-ordered variables is being sampled by each measurement unit (see Table 3.3). While this is true, it is also true that only one taxon — *Lepus* spp. — has a rank order that is totally unaffected by choice of measurement unit, and it is no accident that *Lepus* spp. is by far the most abundant taxon in the collection, ranging from nearly twice to well over twice as abundant as the next-most abundant taxon, depending on the measure selected.

Whether or not these differences in rank orders are of importance to an investigator will depend on the problems being examined. But it should be clear

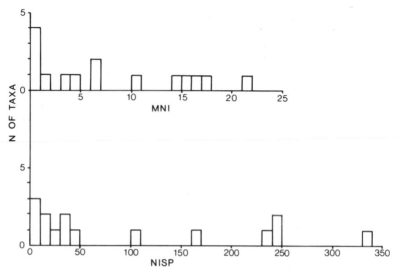

Figure 3.7 The distribution of taxonomic abundances for the Prolonged Drift mammals.

that minimum numbers and specimen counts provide identical rank orders of abundance only in those situations in which the taxa involved are widely separated in abundance. How widely separated they must be is an empirical question whose answer depends on the degree of specimen interdependence and the nature of the distribution of most abundant elements in the fauna involved. There are many faunas in which rank orders based on specimen counts and those based on minimum numbers are very different, even when only the most abundant taxa are examined.

At Prolonged Drift, for instance, the five most abundant taxa as measured by NISP provide rank orders that are only weakly correlated with rank orders based on minimum numbers (Spearman's rho = .30, $p > .20$; see Table 3.4). An investigator who is strongly impressed by the problems introduced by interdependence might wish to follow the minimum number rankings, in which case cattle would represent the most, and Thomson's gazelle the least, abundant taxon of the five. On the other hand, an investigator more impressed with the problems of aggregation might conclude that zebra was the most abundant taxon, while agreeing with the rank of Thomson's gazelle.

Likewise, the 10 most abundant taxa at the Buffalo site are the same whether measured by MNI or NISP, but the rank order of only 1 taxon remains unaltered between the two sets of ranks provided by these units, and the correlation between those ranks is not impressively high ($r_s = .53, .20 > p > .10$; see Table 3.5). There can be no disagreement that deer is the most abundant taxon in this

TABLE 3.2
Rank Orders of Abundance from all Abundance Measures: Connley Cave No. 4 Mammals

Taxon[a]	MNI_{10cm}	$MNI_{stratum}$	MNI_{site}	NISP
1	1	1	1	1
2	2	2	3	2
3	4	5.5	4	3
4	4	4	5	4
5	4	3	2	5
6	8	15	11.5	6
7	6	5.5	6	7
8	8	10	11.5	8
9	8	7	7.5	9
10	11.5	15	18	10
11	10	8	7.5	11
12	14	10	11.5	12.5
13	11.5	15	18	12.5
14	17	15	18	15
15	14	15	18	15
16	14	10	9	15
17	17	15	18	17
18	17	15	11.5	18
19	20.5	20.5	18	20.5
20	20.5	20.5	18	20.5
21	20.5	20.5	18	20.5
22	20.5	20.5	18	20.5

[a] See Table 2.4 for identification of these taxa.

TABLE 3.3
Rank Order of Rearrangments of Ordinal Taxonomic Abundances for all Pairs of Abundance Measures, Connley Cave No. 4 Mammals

Measurement pair	Kendall's tau
MNI_{10cm} with NISP	.947, $p < .001$
$MNI_{stratum}$ with MNI_{site}	.888, $p < .001$
MNI_{10cm} with $MNI_{stratum}$	.817, $p < .001$
MNI_{10cm} with MNI_{site}	.771, $p < .001$
$MNI_{stratum}$ with NISP	.762, $p < .001$
MNI_{site} with NISP	.711, $p < .001$

TABLE 3.4

The Five Most Abundant Mammalian Taxa at Prolonged Drift[a]

Taxon	NISP	Rank	MNI	Rank
Cattle	259	2	22	1
Kongoni	232	4	18	2
Wildebeest	241	3	17	3
Zebra	340	1	16	4
Thomson's gazelle	165	5	15	5

[a] $r_s = 0.30$, $p > .20$.

fauna, but one can argue about the remaining 9 taxa. Even at Fort Ligonier, where the 2 most abundant taxa are represented by 2730 and 1040 specimens respectively, the rank order of these 2 taxa is reversed when minimum numbers are calculated (Table 3.6). Indeed, the correlation between MNI- and NISP-derived rank orders for the 4 most abundant taxa at Fort Ligonier is not significant ($r_s = .60$, $p > .20$). Were cattle or sheep more frequently utilized at Fort Ligonier? Were deer less important than any domestic mammal to the occupants of this site? The answers to both of these questions can be debated by partisans of NISP and MNI.

There are options other than debate between those analysts who prefer specimen counts and those who prefer minimum numbers. One can, for instance, use the approach taken by Guilday (1970: 186) in the Fort Ligonier report and conclude that "qualitatively . . . a fair idea of the relative importance of the various meat animals in the military diet and relative importance of

TABLE 3.5

The Ten Most Abundant Mammalian Taxa at the Buffalo Site[a]

Taxon	NISP	Rank	MNI	Rank
Odocoileus virginianus	52,896	1	746	1
Sciurus carolinensis	481	2	62	3
Procyon lotor	432	3	94	2
Ursus americanus	391	4	14	10
Castor canadensis	321	5	17	7
Canis familiaris	258	6	25	4
Marmota monax	109	7	16	9
Didelphis marsupialis	106	8	22	5
Oryzomys palustris	66	9	17	7
Urocyon cinereoargenteus	58	10	17	7

[a] $r_s = 0.53$, $.20 > p > .10$.

TABLE 3.6

The Four Most Abundant Mammalian Taxa at Fort Ligonier[a]

Taxon	NISP	Rank	MNI	Rank
Box taurus	2730	1	24	2
Ovis aries	1040	2	28	1
Sus scrofa	202	3	6	4
Odocoileus virginianus	150	4	16	3

[a] $r_s = 0.60$, $p > .20$.

hunting can be derived from the collection." Such an approach is fully appropriate, as was Guilday's conclusion that the calculations of meat weights from the minimum numbers of individuals represented at Fort Ligonier was "patently ridiculous." That is, the analyst can step back from any attempt to quantify these data, argue that neither NISP or MNI values provide trustworthy results, and interpret the numbers qualitatively. A conclusion of this sort is unassailable.

Alternatively, an investigator impressed by the difficulties presented by both minimum numbers and specimen counts might wish to launch a detailed taphonomic investigation of the taxa involved to see why the correlation between MNI- and NISP-derived rank orders is so poor, and to see if taphonomic information might reveal which set of rank orders is most likely to reflect the taxonomic composition of the assemblage. In their discussion of the Prolonged Drift assemblage, for instance, Gifford et al. (1980) provide detailed taphonomic assessments of the remains of the more abundant taxa from the site, focusing on the effects of butchering. They argue that the domestic animals represented in this assemblage were slaughtered and butchered on-site, while wild animals were slaughtered and butchered elsewhere and their remains then "schlepped" back to Prolonged Drift. Schlepping, they suggest, could affect relative abundances based on MNI and NISP values, and I note that the differences in rank orders based on these values could be caused by such a phenomenon. Such concerns led Gifford et al. (1980:68) to question the precision implied by minimum numbers: "MNI figures are handy units of comparison, but they should not be . . . used as the basis of such . . . tenuously derived statistics as meat weights and kilocalorie estimates." It is also difficult to disagree with this conclusion.

The differences between rank orders derived from minimum numbers and specimen counts, then, can be treated by handling this information qualitatively, or by conducting a taphonomic analysis of the assemblage involved in order to discover the reason for the discordance, or both. Guilday (1970) and

Gifford et al. (1980) provide examples. Most investigators, however, simply assume the validity of minimum numbers and use those numbers without question. A critical reader must be wary.

At the moment, we have no external way of knowing whether MNI or NISP is providing an accurate assessment of rank order taxonomic abundances for any given faunal assemblage. Accordingly, I have attempted to examine this tissue internally, by comparing rank orders provided by minimum numbers and specimen counts ("maximum" numbers of individuals). If the rank orders provided by the two measures are identical, or statistically equivalent, then it could be concluded that either measure provides an accurate assessment of rank order abundances. I have argued that such similarity can only be expected when taxa are well-separated from one another in terms of both minimum numbers and specimen counts; when this is not the case, either interdependence or aggregation effects, or both, can be expected to result in inaccurate assessments of rank order abundances. Examination of specific faunal assemblages shows that while there are situations in which rank orders remain stable across measuring units, there are also situations in which taxa that are well-separated in abundance on one measure have their rank orders reversed on the other (for instance, the Fort Ligonier cattle and sheep and the Prolonged Drift zebra and cattle). These situations make the interpretation of rank order abundances less straightforward.

If rank orders of taxonomic abundance are invariant across measurement units, there would seem to be little reason not to use them in statistical analysis. Two definitions of *invariant* might by used in this context: (1) that there are no changes whatsoever in rank orders between the most divisive and the least divisive aggregation methods used in minimum number determination, treating NISP as providing the most divisive approach; and, (2) that there are changes in rank order abundance, but these changes are not statistically significant when measured by a rank order correlation coefficient. If the second definition is the one employed, then it should be realized that, as I have noted, changes in rank orders among rare taxa will primarily involve changes in the numbers of ties, not reversals in rank order, for the simple reason that minimum numbers can take on a larger range of values as the number of identified specimens per taxon increases (see Chapter 2). As a result, if a fauna contains many uncommon taxa, a rank order correlation coefficient might well show no significant differences between ranks based on different measurement units even though the rank orders of abundant taxa have changed considerably. This is so because the relative invariance among uncommon taxa can swamp reversals of rank orders among abundant taxa, and the correlation coefficient thus suggest statistical invariance. In this situation, any attempt to statistically determine rank order invariance might best be applied to the most abundant taxa alone.

Working with one's own data, it is simple to test for rank order invariance.

Given that there will be a limited number of ways of aggregating faunal material to test any given hypothesis about that material, one can test for rank order invariance by comparing the rank orders produced by the most divisive approach (NISP) with those produced by the most agglomerative approaches (for instance, MNI_{site} or $MNI_{stratum}$). If rank orders are invariant — that is, do not change at all or do not change significantly — between the most divisive and the most agglomerative aggregation approaches, then the rank orders may be used in further analyses without concern about the effect that the choice of measurement unit may have had on them.

The same approach may be used with published faunas, although here one rarely has the option of recalculating minimum numbers using an approach to aggregation not used by the author. In this setting, the analyst can simply compare NISP-based rank orders with those based on MNI, realizing that the more agglomerative the approach to aggregation in MNI calculation, the stronger will be the support for the assumption of invariance. In an earlier paper (Grayson 1979b), I suggested that an analyst might simply prepare a frequency diagram of taxonomic abundances based on MNI (of the sort illustrated in Figures 3.2 through 3.7), and noted that if most taxa are represented by many specimens or individuals and are widely separated in abundance, the assumption of invariance is supported. While this is true in a relative sense (the assumption is better supported than if most taxa are represented by a few specimens or individuals and are not widely separated in abundance), examples such as Fort Ligonier show that the effects of aggregation may be so severe as to make this approach very misleading.

If rank orders are not invariant across measurement units, several approaches can be taken. One can always treat the taxa as attributes that may either be present or absent (e.g., Grayson 1982; Lyman and Livingston 1983). However, it is also possible that the differences in rank orders are not so great as to alter the outcome of tests in which they are to be used. To see if this is so, calculate the rank orders of taxonomic abundance that result from NISP and from the least divisive approach to aggregation appropriate to the problem at hand (for instance, $MNI_{stratum}$). Then calculate the test statistic for each set of resultant rank orders compared with every other set of rank orders involved in the test. If all probabilities fall below, or if all fall above, the adopted significance level, the problem is resolved. Rank orders are not invariant between measurement units, but this does not matter since the outcome of the test is not altered (see Bradley 1968 for a similar approach to resolving zero differences or tied observations in data sets). If the outcome of the test is altered, then either a qualitative analysis or a detailed taphonomic investigation is called for.

A simple example will help make this approach clear. At Hidden Cave, Strata IV and V provided the two largest samples of identified mammalian specimens. There are a number of reasons why an analyst might wish to compare the faunas of these two strata. Stratum IV, for instance, contains a major archaeological component, whereas Stratum V does not. Did the presence of people have an impact on the abundances of particular taxa of mammals deposited in Hidden Cave? Given that there are good reasons not to treat either NISP or MNI as having provided a ratio level measure here, this question can be addressed through the analysis of rank order abundances.

Rank order abundances for the seven most common mammalian taxa at Hidden Cave, based on both MNI and NISP, are presented in Table 3.7. It is easy to demonstrate that these rank orders are affected by choice of measurement unit. Within Stratum IV, NISP- and MNI-based rank orders are negatively correlated, but the correlation is not significant ($r_s = -.65, p > .10$). Within Stratum V, NISP- and MNI-based rank orders are positively correlated, but the correlation is not significant ($r_s = +.55, p > .10$). The outcome of comparisons of rank orders between Strata IV and V is seriously affected by these within-stratum changes. An NISP-based comparison would provide a Spearman's rho value of .89 ($p = .01$), and would lead to the conclusion that there had been no significant impact of people on the rank orders of the most abundant mammals deposited in the cave. An MNI-based comparison would provide a Spearman's rho of .35 ($p > .20$), and would lead us to conclude that people did alter the abundances of taxa introduced into the cave, with the greatest alterations involving, in decreasing order of magnitude, *Microtus* sp., *Neotoma cinerea*, and *Sylvilagus nuttallii*. The differences between the NISP- and MNI-based results are serious ones; changes in rank order provided by MNI and NISP support totally different conclusions. I could now either analyze these data qualitatively, or begin a taphonomic analysis. Were I to do a taphonomic study, I would begin by focusing on those taxa whose rank order abundance are most seriously affected by the shift from specimen counts to minimum numbers as a measuring unit. As Table 3.7 shows, the most seriously affected taxon is *Lepus* sp.; taphonomic analysis would quickly show that *Lepus* remains are more highly fragmented than those of any other abundant taxon, and that the shift in rank — from first on the NISP list to fifth or sixth on the MNI list — could have been expected from this fact alone. These and similar considerations might lead to a reasoned choice of rank orders to be used in any analysis of changing abundances within Hidden Cave. Without such a procedure, however, our conclusions would depend on a less informed choice between NISP and MNI as a counting unit. For the reasons I have already discussed, were I to make a less informed choice, I would choose NISP-based ranks.

There are, then, situations in which specimen counts and minimum numbers can provide valid ordinal scale information on taxonomic abundances. But there are also situations in which rank orders are severely affected by choice of measurement unit, including the rank orders of taxa well-represented and well-separated in terms of specimen counts and minimum numbers. I have

TABLE 3.7
NISP- and MNI-Based Rank Order Abundances for the Seven Most Abundant Mammalian Taxa at Hidden Cave[a]

Taxon	NISP				MNI			
	Stratum IV	Rank	Stratum V	Rank	Stratum IV	Rank	Stratum V	Rank
Lepus sp.[b]	429	1	255	1	32	6	5	5
Sylvilagus sp.[c]	230	2	227	2	37	4.5	12	2.5
Marmota flaviventris	140	4	68	6	12	7	3	7
Dipodomys spp.[d]	113	5	99	4	59	1	12	2.5
Neotoma lepida[e]	76	6	86	5	38	3	8	4
Neotoma cinerea[f]	142	3	149	3	37	4.5	13	1
Microtus sp.[g]	47	7	24	7	42	2	4	6

[a] Spearman's rho and associated probabilities: Stratum IV, NISP–MNI: −0.65, $p > .10$; Stratum V, NISP–MNI: +0.55, $p > .10$; Strata IV–V, NISP: +0.89, $p = .01$; Strata IV–V, MNI: +0.35, $p > .20$.
[b] Includes *L. californicus*.
[c] Includes *S.* cf. *nuttallii*.
[d] Includes all species of *Dipodomys*.
[e] Includes *N.* cf. *lepida*.
[f] Includes *N.* cf. *cinerea*.
[g] Includes *M. montanus*.

suggested ways of resolving the difficulties posed by the latter situation, and now wish to return to an argument I made in Chapter 2. There, I maintained that specimen counts contain much of the information on taxonomic abundance contained in minimum numbers. Now, however, I have argued that MNI and NISP may give very different accounts of rank order abundance. Are these statements not in conflict?

The answer depends on the scale at which the relationships are examined. At Prolonged Drift, for instance, $\log_{10}$ MNI and $\log_{10}$ NISP are highly correlated across all taxa ($r = .93$, $p < .001$), as are MNI- and NISP-based rank orders of taxonomic abundance ($r_s = .92$, $p < .001$). However, $\log_{10}$ MNI and $\log_{10}$ NISP may or may not be highly correlated for any selected subset of those taxa. As I have discussed, MNI- and NISP-based rank orders for the five most abundant taxa at Prolonged Drift are not significantly correlated ($r_s = .30$, $p > .20$), nor are the abundances of those taxa as measured by $\log_{10}$ MNI and $\log_{10}$ NISP ($r = .20$, $p > .50$). The relationship between $\log_{10}$ MNI and $\log_{10}$ NISP for the five most abundant Prolonged Drift taxa is shown in Figure 3.8, where it is obvious that while cattle are represented by more individuals than zebra, zebra are represented by more specimens than cattle. At Prolonged Drift, specimen counts contain much the same information on abundance as is contained in minimum numbers across all taxa, but this is not true for selected subsets of those taxa. This situation also characterizes many of the other sites I discussed in Chapter 2.

As regards entire faunal assemblages, then, specimen counts contain much the same information on taxonomic abundance as minimum numbers contain, while not registering the effects of aggregation. However, NISP and MNI may provide very different suggestions concerning the abundances of taxonomic subsets of those assemblages. To resolve the differences in information on abundances provided by these two measures requires detailed taphonomic data of a sort that may or may not be readily available, or requires that an analyst rely heavily on qualitative assessments of taxonomic abundance. An analyst as impressed as I am with the problems introduced by aggregation might wish to adhere more closely to specimen count data, but it should be clear by now that such a solution is not without serious problems of its own.

Conclusions

Where are we left? Specimen counts and minimum numbers can rarely provide unambiguous ratio scale measures of taxonomic abundance. If, perchance, either unit should have provided a ratio scale measure for any given assemblage, we have no way of knowing it. Either specimen counts or minimum numbers may provide a valid ordinal scale measure of taxonomic abun-

CONCLUSIONS

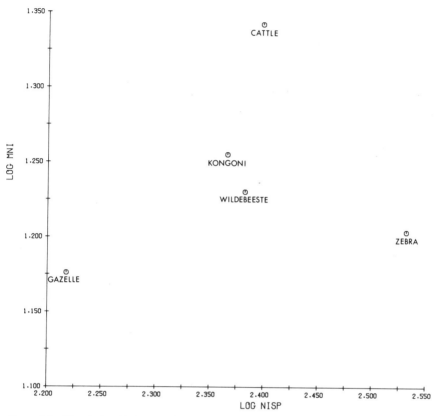

Figure 3.8 The relationship between $\log_{10}$ MNI and $\log_{10}$ NISP for the five most abundant taxa at Prolonged Drift.

dance, and there are cases in which empirical support can be provided for the argument that such a scale has been achieved. There are other cases, however, for which empirical support is not possible, and resolution in these cases must come through detailed taphonomic analysis if it is to come at all. If an analysis of this sort cannot be done, then qualitative assessment of taxonomic abundances, as done by Guilday (1970) with the Fort Ligonier material, is fully appropriate.

All these critiques, however, apply only to faunas from single sites. When faunas from many sites in the same region are available (and the definition of "region" depends on the problem being addressed), cross-checking sets of taxonomic abundances become available as well. Such sets can provide a powerful means of addressing the validity of any single set, whether the target of analysis be human subsistence or past environments. When, for instance, changes in taxonomic abundance through time at a single site are matched by changes through the same period of time at other sites in the same region, it is reasonable to conclude that changing taxonomic abundances are, in fact, being accurately measured.

A single example will suffice. At the Connley Caves, pikas *(Ochotona princeps)* occurred in Strata 3 and 4, dated to circa 11,000 to 7000 B.P., but not in Strata 1 and 2, dated to approximately 4400 to 3000 B.P. In the Great Basin of the arid western United States, pikas now occur as isolated populations on mountains, but do not exist in the intervening lowlands, which are too hot and/or dry for them. Although pikas have been reported in the highlands to the west, south, and east of the Fort Rock Basin in which the Connley Caves are located, there are no records for them in the basin itself. It would appear that, although sufficient talus for pika habitat is present here today, sufficient succulent vegetation is lacking. In short, the modern Fort Rock Basin appears to be too xeric for these animals. Between 7000 and 11,000 years ago, however, the animals were clearly present in the area. The number of pika specimens from the caves is so small (Table 3.8) that few if any analysts would place faith in them other than as indicative of the fact that during the early Holocene, pikas lived close enough to the Connley Caves to have become incorporated in the deposits that fill those sites.

The timing of the disappearance of pikas from this area, however, is more problematic. First, there is a gap of approximately 2500 years between the top of Stratum 3 (marked by the deposition of Mazama ash, which resulted from the eruption of Mt. Mazama some 100 km to the west of the caves) and the bottom of Stratum 2. Second, the absence of pika specimens in Strata 1 and 2 does not mean that they did not live in the Fort Rock Basin at this time. It means only that if they did live here, their bones did not end up in the Connley Caves, or, if their bones were deposited in these sites, they were not preserved, or, if they were preserved, the bones were not collected. In addition, only 21% of the total number of identified mammalian specimens from the Connley Caves came from Strata 1 and 2 (Grayson 1979a) and, as I discuss in Chapters 4 and 5, large samples are needed to detect rare animals. As a result, even if pikas were still living in the area during the times the uppermost strata of the Connley Caves accumulated, the post-Mazama decrease in sample size within these caves could have caused their absence in the uppermost strata. One can accept as a working hypothesis that pikas were extirpated from the Fort Rock Basin between 7000 and 4500 B.P., but the Connley Caves provide little more than this speculation concerning the timing of their local extirpation. In addition, if this hypothesis is correct, there are two plausible explanations for the local loss of pikas. First, it is possible that pikas succumbed to increasing temperatures and/or decreasing effective precipitation sometime after 7000 B.P., a climatic event that is well documented for many parts of the arid west (Spaulding *et al.*

CONCLUSIONS

TABLE 3.8

Minimum Numbers of Individuals (MNI_{10cm}) and Numbers of Identified Specimens (NISP) of Pika and All Other Mammals by Stratum within Connley Caves 3, 4, 5, and 6[a]

Stratum	Pika		All other mammals	
	MNI	NISP	MNI	NISP
1	0	0	73	122
2	0	0	66	155
3	5	10	174	718
4	6	6	113	541
	11	16	426	1536

[a] From Grayson (1979a).

1983), and that conditions have been too xeric since that time to allow recolonization. It is also possible that the deposition of Mazama ash in the Fort Rock Basin at ca. 7000 B.P. adversely impacted these animals by filling the talus on which they depend for shelter, by decimating the local vegetation on which they depend for food, and by accelerating toothwear through the abrasive effects of masticated pumice. An overlapping, or even subsequent, shift towards increasingly xeric conditions might then have prevented recolonization of the Fort Rock Basin by these animals (see the discussions in Grayson 1977a, 1979a). The Connley Caves fauna, then, leaves us with many questions.

Gatecliff Shelter also provided a record for pikas. Although pikas are still present in the Toquima Range, the lowest published records for them are some 300 m above the elevation of Gatecliff Shelter (Hall 1946). Of the 58 pika specimens present in the Gatecliff fauna, 95% of them (55 specimens) were deposited prior to 5000 years ago; during the last 5000 years, 73% (9507 specimens) of the Gatecliff small mammal specimens were deposited, but this fauna includes only three pika bones. There are, of course, many reasons why this shift in pika abundance might have occurred. It is possible that the number of pikas in the area remained unchanged, while talus slopes near Gatecliff became choked with sediment, removing pika habitat in the vicinity of the site. It is also possible that, for some unknown reason, accumulation mechanisms simply stopped sampling these animals, even though their abundances were not locally diminished. Certainly, the distribution of pika specimens across strata within Gatecliff does not suggest a slow decline in deposition through time (Table 3.9; the raw data are presented in Table 1.1).

TABLE 3.9
Numbers of Identified Pika Specimens by Grouped Strata within Gatecliff Shelter

Strata	Years B.P.	NISP, Pika	Stratum NISP	% Pika
1	0–1250	2	3519	00
2–5	1250–3200	0	3731	00
6–19	3200–5000	1	2139	00
20–37	5000–6250	30	1824	02
54–56	6300–ca. 7300	25	1685	01

If Gatecliff stood alone in the arid west, the decreasing numbers of pika specimens through time at this site would support few convincing arguments about the history of pikas in the Great Basin. It could be hypothesized that these decreasing numbers reflect an increase in the lower elevational limits of pikas in the Toquima Range (Grayson 1983b), but the poor control we have over the meaning of specimen counts would force us to stop there. The Connley Caves, however, show what appears to be an identical phenomenon, differing only in that pikas have been extirpated from the Fort Rock Basin while they continue to exist in the Toquima Range. That similar declines occur in faunas that accumulated some 600 km apart lends credence to the numbers provided by both. In addition, woodrat middens from the Snake Range in eastern Nevada provide a very different kind of faunal record for pikas, one based on coprolites, that provides a very similar picture. Here, woodrats were incorporating pika coprolites into middens on the lower reaches of the Snake Range between 17,000 and 6000 B.P., but not after that time. While the numbers provided by any one of these faunal sequences can be reasonably questioned, the consistency provided by all cannot. Indeed, the late Pleistocene and Holocene history of pikas has found a major place in a general theoretical model of the biogeographic history of boreal mammals in the Great Basin (Brown 1971, 1978; Grayson 1977a, 1981a, 1982, 1983b; Mead et al. 1982; Spaulding et al. 1983; Thompson and Mead 1982).

The example I have just provided focuses on the analysis of past environments in general, and on mammalian biogeography in particular. Many other paleoenvironmental examples exist. John Guilday and his colleagues, for instance, have demonstrated impressive correlated changes in the abundances of small mammals during the past 15,000 years in a series of sites scattered across a large portion of the eastern United States (Guilday et al. 1978). Although one might have reasonably doubted the changing minimum numbers per taxon provided by any one of these sites taken separately, there is little reason to question the validity of the temporally correlated shifts in abundance docu-

CONCLUSIONS

mented by the entire set. Subsistence-oriented analyses can provide similar examples. There are, for instance, impressive correlations in the rank orders of taxonomic abundances of the more common mammals present in a wide variety of prehistoric open sites in New York state, and the similarity of these rank orders suggests that they are providing valid information on prehistoric human subsistence in this area (see the general discussion in Grayson 1974b).

What if faunal assemblages cannot be inspected for redundancies in taxonomic abundances? After all, there are many regions whose faunal history is known only from a single site. Hidden Cave, for instance, provides the only mammalian sequence for the southern Carson Desert of Nevada that spans late Pleistocene and Holocene times, and this sequence simply cannot be compared to other sequences in the region at the current time. In the subsistence setting, interassemblage variability, not interassemblage similarity, is often the focus of analysis, and in such cases comparisons of taxonomic abundances cannot be used to increase our faith in the validity of the abundances provided by any given site. Here, it would seem that the analyst must employ internal analyses of the kind I have been discussing. Are the rank orders of taxonomic abundance invariable when measured by both NISP and MNI? If not, can detailed taphonomic studies lead to a reasoned decision as to which of these sets of rank orders is most likely measuring what the analyst is attempting to measure? In those cases in which taphonomic studies cannot be done, or in which such studies do not provide an answer, the analyst can either fall back on qualitative assessments, or, for the reasons that I discussed in Chapter 2, can simply use specimen counts as the basis of his or her quantitative analyses.

CHAPTER 4

Sample Size and Relative Abundance

Unless a paleontological or archaeological site is completely excavated and every bone that was deposited in that site retrieved, the faunal assemblage recovered is a sample of the entire set of bones that could have been recovered. The entire set of bones that could have been retrieved is, in turn, a sample of some other population, the nature of which depends on the kinds of questions being asked. In the paleoenvironmental setting, for instance, the target population is usually the set of animals, and the relative abundances of those animals, that existed in the area surrounding the site at the time the fauna accumulated. In the subsistence setting, the target population is often the set of animals, and the relative abundances of those animals, that were utilized by a group of people (e.g., Smith 1979). In one way or another, then, a faunal analyst is always working with samples. The only real direct control the analyst has over the nature of those samples lies in the means used to retrieve them from the ground. Control over the relationship between the retrieved sample and the deposited sample, and between the deposited sample and the target population, must routinely be indirect (see the discussions in Gifford 1980; Grayson 1979b; Lyman 1982a; and references therein).

Because of the poor control faunal analysts generally have over the relationship between their retrieved sample and the target population, there are many ways in which the relative abundances within that sample can become biased estimators of relative abundances in the population. I have already discussed some of these ways; the taphonomic literature contains insightful discussions of many others (e.g., Binford 1981; Brain 1981; Gifford 1981; Gilbert and Singer 1982; Lyman 1982a). Here, I explore a statistical artifact that can result from working with poorly controlled samples: significant correlations between the size of our samples and the relative abundances of the taxa that comprise those samples.

All faunal analysts recognize that extremely small samples probably do not provide an adequate base from which statistical inferences concerning relative abundances within the target population can be made. However, it is also true that because we have so little control over the relationship between the retrieved sample and the target population, faunal analysts generally, and understandably, make the facilitating assumption that if samples are large enough, relative abundances may be extracted and manipulated without serious concern over sample size effects. In practice, samples that intuitively appear to be "too small" are rejected as the basis for statistical manipulation, while those that appear to be "large enough" are so used. This is true for the simple reason that very little is known about how large a sample must be to adequately represent a population in any given instance. Routinely, the statistical manipulations involved in these studies begin with, and often end with, the determination of relative taxonomic abundances using NISP or MNI.

Analysis of studies that utilize relative abundance data suggests that the changing relative abundances that have been detected by these studies may often not be reflecting the different values of the parameters of interest, but may instead be reflecting the differing sizes of the samples from which the relative abundances have been derived. In almost no case is the effect obvious, and the initial examples I present here are not meant to be critical of the scientists whose work I analyze. Indeed, it is the very subtlety of the effect that makes it so pernicious: there is often no obvious reason to expect the problem to occur and hence no reason to look for it, unless one is aware of the issue beforehand. Here, I begin by presenting some simple examples in order to illustrate the effect, and then present lengthier examples in order to explore the reasons for it.

Correlations between Sample Size and Relative Abundance: Some Examples

Example 1: Snaketown, Southern Arizona

Included in the classic study of the prehistoric Hohokam village of Snaketown (Haury 1976) is a brief analysis of a sample of the vertebrate fauna from the site (Greene and Matthews 1976). Greene and Matthews provide a figure that displays the frequency of "the principal meat-producing animals by phase" (1976:371), calculated as the percentage of the minimum number of individuals per taxon by Hohokam phase. The abundances of the six taxa included in this figure vary from phase to phase, although Haury (1976) notes that these variations do not seem culturally significant. In fact, analysis of the numbers provided by Greene and Matthews suggests that they might not be telling us about subsistence at all.

TABLE 4.1

Minimum Numbers of Individuals and Relative Abundances of Deer by Phase at Snaketown[a,b]

Phase	MNI	Rank, MNI	% Deer	Rank, % deer
Vahki	65	1	15.4	6
Estrella	15	6	40.0	2
Sweetwater	56	2	25.1	4
Snaketown	27	4	11.1	7
Gila Butte	49	3	24.5	5
Santa Cruz	23	5	39.1	3
Sacaton	4	7	50.0	1

[a] From Greene and Matthews (1976).
[b] Spearman's rho, MNI-% deer, $= -0.75$ ($p < 0.10$).

Table 4.1 presents minimum numbers of individuals for all mammals by unmixed phase at Snaketown, and also presents the relative abundance of deer expressed as a percentage of the total MNI for each phase. These two sets of numbers suggest that something other than the changing relative abundances of deer may be being measured here. As sample size increases, the relative abundance of deer seems to decrease. Spearman's rho confirms this inverse relationship: $r_s = -.75$ ($p < .10$). I note that the only phases represented in Table 4.1 are those displayed in the figure provided by Greene and Matthews (1976:371); addition of the Pioneer phase to the analysis does not alter the results ($r_s = -.74$, $p < .10$).

As a result, it is no longer clear whether these relative abundances are measuring changing abundances of deer through time, or are instead being determined by differing sample sizes across phases. The significant negative correlation between sample size and the relative abundance of deer suggests that these relative abundances may not be a valid measure of the relative importance of deer characteristic of Snaketown phases.

Example 2: Hogup Cave, Northwestern Utah

Neither Haury (1976) nor Greene and Matthews (1976) based further analyses on the changing relative abundances of deer, and most other taxa, across Snaketown phases. As a result, the fact that these changing abundances may be a function of changing sample sizes does not harm their interpretation of the Snaketown fauna. It is not hard, however, to find situations in which possible sample size effects on relative abundances render the interpretation of those abundances much less convincing.

An example can be drawn from the work of Harper and Alder (1970). Harper

TABLE 4.2

Sample Sizes and Relative Abundance of Xeric Rodents by Major Stratigraphic Unit, Hogup Cave[a,b,c]

Hogup unit	MNI	Rank	Xeric plus mesic rodents, MNI	Rank	% Xeric rodents	Rank
1	444	2	135	2	43	5
2	2247	1	371	1	61	4
3	298	4	79	3	75	3
4	303	3	44	4	89	2
5	84	5	29	5	93	1

[a] From Harper and Alder (1970).
[b] Spearman's rho, MNI-% xeric rodents = -0.80 ($p = 0.10$).
[c] Spearman's rho, xeric plus mesic rodents-% xeric rodents = -0.90 ($p = 0.05$).

and Alder presented a detailed and thoughtful paleoenvironmental analysis of the plant and animal remains from Hogup Cave, northwestern Utah. In order to monitor changing abundances of mesic and xeric habitats through time in the area surrounding Hogup Cave, Harper and Alder divided the rodents from the site into mesic and xeric groups, eliminating from consideration those that contributed less than one individual per 1000 years to the deposits. They then plotted changing relative abundances of mesic and xeric rodents against the five major Hogup stratigraphic units, and demonstrated that the relative abundance of xeric rodents climbed steadily from the time of deposition of the earliest unit, between 6400 and 1250 B.C., to the time of deposition of the latest unit, between A.D. 1350 and 1850. They concluded that the "relative abundance of mesic and xeric forms would seem to suggest continued drying of the upland environment throughout the period of time [represented in the deposits]" (Harper and Alder 1970:234).

Table 4.2 presents total numbers of mammalian individuals per Hogup Cave stratigraphic unit, as well as the percentage of the rodent fauna assigned by Harper and Alder (1970) to the xeric group. Simple inspection of this table shows that as the total MNI for a stratigraphic unit increases, the relative abundance of xeric rodents decreases; Spearman's rho for this relationship is $-.80$ ($p = .10$). Table 4.2 also presents the total number of rodents assigned to Harper and Alder's mesic and xeric categories; Spearman's rho for the relationship between the total number of rodents analyzed and the percentage of xeric rodents is $-.90$ ($p = .05$).

As with the Snaketown data, these significant correlations between sample size and relative abundances make it reasonable to question the assertion that "percentage xeric rodents" is measuring the abundance of xeric habitats in the area surrounding Hogup Cave. It appears that this figure may be largely a function of sample size: as sample size increases, the relative abundance of xeric rodents decreases. It is not unreasonable, as a result, to suggest that in this instance "percentage xeric rodents" may not be a valid measure of the abundance of xeric habitats surrounding Hogup Cave in the past.

Example 3: Raddatz Rockshelter, South-central Wisconsin

In 1966, Cleland published a significant study of the prehistoric animal ecology and prehistoric human use of vertebrates in the upper Great Lakes region. As part of this analysis, he re-examined the vertebrate remains from Raddatz Rockshelter (Wittry 1959), which had previously been studied by Parmalee (1959). While Parmalee (1959) had interpreted the abundances of the taxa from this site on essentially nominal and ordinal levels, Cleland (1966) wished to make more detailed statements about changing environments and changing human utilization of the prehistoric fauna through time. Of interest here is his attempt to infer changing habitat types through time from changes in percent abundance of a set of taxa that he felt were indicative of those habitat types.

Table 4.3 presents the total number of identified specimens per level at Raddatz Rockshelter, the percentage of those specimens that are deer, and the

TABLE 4.3

Sample Sizes and Relative Abundances of Deer, Raddatz Rockshelter[a,b]

Level	Total NISP	Rank	% Deer	Rank
1	117	11	93 (93.16)	5
2	181	7	90	6
3	286	6	87	7
4	374	4	95	2
5	551	2	96	1
6	575	1	93 (93.16)	4
7	442	3	94	3
8	344	5	86	8
9	179	8	80	9
10	170	9	69 (69.82)	11
11	125	10	63	12
12	82	12	56	13
13	49	14	69 (69.39)	10
14	69	13	45	14
15	44	15	43	15

[a] From Cleland (1966). Values in parentheses are unrounded figures.
[b] Spearman's rho, NISP–% deer = 0.84 ($p < 0.001$).

rank orders of the levels in terms of both total NISP values and percentage deer. I have analyzed deer in this example because that taxon is the most abundant taxon in every level. Because of the interdependence of percentage values, the behavior of other taxa will tend to be strongly correlated with that of deer (see Chapter 2). As Table 4.3 shows, Spearman's rho between NISP values per level and percentage deer per level is very high (r_s = .84, p < .001). As a result, it becomes difficult to interpret changing relative abundances through time at this site. Perhaps, as Cleland (1966) suggested, they are providing information on changing abundances of habitat types through time in the region surrounding Raddatz Rockshelter. Perhaps, however, they are primarily telling us about the number of bone specimens identified per level.

Exploring the Causes

In each of the examples I have presented, there is a significant correlation between the relative abundances of taxa, or of composite groups composed of several taxa, and the size of the samples from which those relative abundances were defined. There are, of course, two possible general reasons for these correlations. First, it is possible that no causal relationship exists between sample size and relative abundance in these instances; some third factor, for instance, may be causing both sets of changes. In the case of Hogup Cave, it is possible that both increasing abundances of xeric rodents and decreasing sample sizes are caused by an increasingly arid environment. Were this the case, then the relative abundances of xeric rodents might, in fact, provide a good indicator of aridity, as might the sample sizes themselves. Second, changing sample sizes might be determining relative abundances, in which case those relative abundances cannot be used as a valid measure of the parameter of interest unless, perchance, the sample sizes themselves reflect that parameter. It is clear, however, that the mere presence of such a correlation greatly clouds any interpretive use that might be made of the relative abundances involved.

In most cases, the cause of the correlation is clear. Such cases are provided by situations in which a number of strata have very small NISP values; removing those strata from analysis also removes the significant correlation.

Gatecliff Shelter provides a number of cases of this sort. For instance, I examined the changing relative abundances of pygmy rabbits *(Sylvilagus idahoensis)* through time at Gatecliff (Grayson 1983b) to see if their numbers here showed a mid-Holocene decline similar to that which I had detected at the Connley Caves (Grayson 1979a) and which Butler (1972) had detected at the Wasden site, southern Idaho. Inspection of the relative abundances of *S. idahoensis* across all Gatecliff strata shows that while the lower faunal assemblages tend to be characterized by higher relative abundances of pygmy rab-

TABLE 4.4

Relative Abundances of *Sylvilagus idahoensis* (*S. idahoensis* and *S.* cf. *idahoensis*) through Time at Gatecliff Shelter: All Strata

Stratum	NISP (*S. idahoensis*)	Relative abundance (NISP/total stratum NISP)
1	86	.02
2	4	.02
3–5	63	.02
6/7	6	.02
8	7	.09
9	10	.02
10	0	.00
11/12	23	.04
13	3	.01
14–16	0	.00
17	0	.00
18	0	.00
19	7	.02
20	2	.04
21	0	.00
22	8	.09
23	0	.00
24/25	47	.11
26–30	0	.00
31/32	35	.14
33	41	.07
37	63	.23
54	69	.14
55	0	.00
57	268	.22

bits, there are also great fluctuations in relative abundance from stratum to stratum, fluctuations that do not match the changes seen at the Connley Caves and the Wasden site. However, there is also a significant correlation between total stratum NISP values and the percentage of those specimens that are *S. idahoensis* at Gatecliff ($r_s = .60$, $p < .01$; see Table 4.4). This correlation is due to the effects of strata with very small total numbers of identified specimens. Accordingly, removing strata with less than 150 identified specimens from the analysis also removes the significant correlation ($r_s = .01$, $p > .50$; see Table 4.5). Inspection of pygmy rabbit relative abundances in those strata with more than 150 identified small mammal specimens reveals a rather straightforward pattern of decreasing pygmy rabbit abundance through time in the Gatecliff fauna. Relative abundances of 7% and higher characterize strata deposited prior

TABLE 4.5

Relative Abundances of *Sylvilagus idahoensis* (*S. idahoensis* and *S.* cf. *idahoensis*) at Gatecliff Shelter: Strata with More Than 150 Total Identified Specimens

Stratum	NISP (*S. idahoensis*)	Relative abundance (NISP/total stratum NISP)
1	86	.02
2	4	.02
3-5	83	.02
6/7	6	.02
9	10	.02
11/12	23	.04
13	3	.01
19	7	.02
24/25	47	.11
31/32	35	.14
33	41	.07
37	63	.23
54	69	.14
56	268	.22

to ca. 5200 B.P., while relative abundances of 4% and lower characterize strata deposited after this time. While these single site abundances might or might not be meaningful in-and-of themselves (and the analysis of them that I have presented elsewhere is essentially qualitative [Grayson 1983b]), they are fully in line with the information provided by the Connley Caves and by the Wasden site. As a result, it is reasonable to argue that all three sites together reflect the regional Holocene history of *S. idahoensis* in the northern Great Basin and immediately adjacent southern Plateau (see the overview in Grayson 1982).

The Connley Caves also revealed a large decrease in the abundance of *Lepus* spp. after 7000 B.P., a decrease that I noted but did not attempt to explain (Grayson 1977a, 1979a). Across all Gatecliff strata, the relative abundance of *Lepus* seems generally higher in deeper levels and generally lower in upper ones, but certainly no clear pattern of change similar to that presented by the Connley Caves emerges. The correlation r_s between total stratum NISP and the relative abundance of *Lepus* (*Lepus* sp., *Lepus* cf. *townsendii*, and *Lepus* cf. *californicus*) in each stratum is .33 ($p = .11$; see Table 4.6). Removing strata with less than 150 total identified small mammal specimens reduces this coefficient to .18 ($p > .50$; see Table 4.7). The pattern now becomes clearer. The most recent stratum with more than 150 identified small mammal specimens and with relative abundances of *Lepus* of greater than 5% is 24/25 (ca. 5300 B.P.) With the exception of Stratum 33, strata earlier than 24/25 are marked by

TABLE 4.6

Relative Abundances of Lepus (*Lepus* sp., *L* cf. *townsendii*, and *L.* cf. *californicus*) through Time at Gatecliff Shelter: All Strata

Stratum	NISP (*Lepus*)	Relative abundance (NISP/total stratum NISP)
1	134	.04
2	6	.04
3–5	121	.03
6/7	4	.02
8	2	.03
9	32	.05
10	17	.59
11/12	0	.00
13	1	.00
14–16	0	.00
17	1	.02
18	0	.00
19	4	.01
20	3	.07
21	0	.00
22	10	.11
23	5	.12
24/25	33	.08
26–30	0	.00
31/32	25	.10
33	28	.05
37	37	.14
54	67	.14
55	0	.00
57	186	.15

yet higher abundances of hares. The Gatecliff record thus seems to match that provided by the Connley Caves: more specimens of *Lepus* were being deposited during the earlier Holocene than during later times (see also Grayson 1977a, 1982).

Hidden Cave provides examples as well. Marmots *(Marmota flaviventris)*, for instance, have a complex and fascinating history in the arid west (Grayson 1982, and references therein; Harris 1977). What is known of that history suggests that these animals should have been more abundant during the cooler and/or moister late Pleistocene and early Holocene (corresponding to Hidden Cave Strata XIV to VII) in the area surrounding Hidden Cave than after that time. The botanical studies conducted by Wigand and Mehringer (1984) also suggest that such a shift should have occurred here: their analysis implies the onset of

TABLE 4.7
Relative Abundances of *Lepus* (*Lepus* sp, *L.* cf. *townsendii*, and *L.* cf. *californicus*) through Time at Gatecliff Shelter: Strata with More Than 150 Total Identified Specimens

Stratum	NISP (*Lepus*)	Relative abundance (NISP/total stratum NISP)
1	134	.04
2	6	.04
3–5	121	.03
6/7	4	.02
9	32	.05
11/12	0	.00
13	1	.00
19	4	.01
24/25	33	.08
31/32	25	.10
33	28	.05
37	37	.14
54	67	.14
56	186	.15

warmer and/or drier conditions, and the establishment of vegetation resembling that of modern times, soon after Stratum VII was deposited.

Table 4.8 presents the number of identified specimens of *Marmota flaviventris* per Hidden Cave stratum, the total number of identified mammal specimens for each stratum, and the percentages of those specimens that are marmot. Inspection of this table shows that the changing relative abundances of marmots within Hidden Cave do not match expectations derived from Wigand and Mehringer's work and from general knowledge of marmot history in the arid west. Several transitions from stratum to stratum are marked by major and alternating increases and decreases in marmot relative abundance, and Stratum IV (3700–3800 B.P.) has a relative abundance of marmots far greater than expected. However, it is also true that the rank order of relative abundances of marmots is significantly correlated with the rank order of stratum NISP values ($r_s = .65$, $p < .05$). This correlation is caused by small-sample strata, and can be eliminated by removing from analysis those strata with a total of fewer than 145 identified specimens ($r_s = -.21$, $p > .20$; see Table 4.9). Unfortunately, this approach also eliminates the possibility of examining two of the three late Pleistocene strata (XIV and XI). With the sample size correlation removed, it appears that Strata XIII (21,000–18,000 B.P.), IX (10,000–7500 B.P.), VII

TABLE 4.8

Relative Abundances of *Marmota flaviventris* Through Time at Hidden Cave: All Strata with Identified Mammalian Specimens

	NISP		
Stratum	*Marmota*	All mammals	% *Marmota*
I	5	212	02
II	10	286	04
III	8	189	04
IV	140	1374	10
V	68	1069	06
VII	27	428	06
VIII	0	1	00
IX	12	158	08
XI	0	8	00
XIII	10	145	07
XIV	0	7	00

(7500–6900 B.P.), and V (6900–5400 B.P.) have essentially identical abundances of marmots, while the three most recent strata (3700–0 B.P.) are marked by lower relative abundances of these animals. Given that other paleoenvironmental data suggest that the major decrease in marmot abundance here should have occurred at roughly 6900 B.P., these results, including the high relative abundance of marmots in Stratum IV (3800–3700 B.P.), are not in good accord with our expectations. The possible meaning of these results may not become clear until faunal sequences from other parts of the southern Lahontan Basin are available. At least, however, we can now be sure that the changing relative abundances of marmots within Hidden Cave that have provided these results are not a function of changing sample sizes (for a more detailed discusson of these data, including an analysis that incorporates all late Pleistocene strata by creating stratigraphically composite samples using data from mixed strata, see Grayson 1984).

In the Gatecliff and Hidden Cave examples I have just discussed, and in most cases that I have been able to examine in detail, correlations between sample size and relative abundances are caused by strata with very small faunal assemblages. The cause is an obvious one (although Gatecliff and Hidden Cave provide particularly obvious instances), and the relationship can be removed by dropping strata with small samples from the analysis. Given that we know relatively little about the relationship between sample size and relative abundance, the approach I have used to solve this problem is completely empirical. I rank order assemblages according to their NISP values and remove assem-

EXPLORING THE CAUSES

TABLE 4.9

Relative Abundances of *Marmota flaviventris* Through Time at Hidden Cave: Strata with 145 or more Identified Mammalian Specimens

Stratum	NISP		% *Marmota*
	Marmota	All mammals	
I	5	212	02
II	10	286	04
III	8	189	04
IV	140	1374	10
V	68	1069	06
VII	27	428	06
IX	12	158	08
XIII	10	145	07

blages in order of increasing sample size until the correlation between sample size and relative abundance is no longer significant. This approach is not elegant, but it works. Hopefully, fuller understanding of these relationships will lead to more elegant solutions.

Finally, I return to Raddatz Rockshelter, since the cause of the correlation between sample size and relative abundance here is of a very different sort than that discussed above. Recall that within Raddatz Rockshelter numbers of identified specimens per stratum and the percentage of those specimens that are deer are significantly correlated ($r_s = .84$, $p < .001$; see Table 4.3). I concluded that changing relative abundances of deer at this site might be telling us about changing habitat types (accepting Cleland's assumption that habitat change was the cause of changing abundances of deer here), or about changes in the number of bone specimens per level.

Removing levels with small samples from the Raddatz analysis does not remove the correlation. Dropping levels with total NISP counts of less than 150, for instance, only reduces the correlation (rho) to .82 ($p < .001$). Fortunately, in this case it is easy to demonstrate the cause of the significant correlation, and to explain why it cannot be removed by dropping small-sample levels. Once this cause is understood, it can be taken into account in any interpretation of the fauna from this site.

Table 4.10 presents the number of identified deer bones per level at Raddatz Rockshelter, as well as the number of identified specimens of all other vertebrates. Note that while deer NISP values vary from 19 (Level 15) to 536 (Level 6), the number of all other vertebrate specimens varies only from 8 (Level 1) to 53 (Level 10). The largest deer NISP value is 28.2 times larger than the smallest deer NISP value, but the largest NISP for all other taxa is only 6.6 times larger

TABLE 4.10

The Number of Identified Deer Bones and of All Other Vertebrates by Level at Raddatz Rockshelter

Level	NISP		
	Deer	All other vertebrates	Total
1	109	8	117
2	163	18	181
3	250	36	286
4	354	20	374
5	527	24	551
6	536	39	575
7	414	28	442
8	296	48	344
9	144	35	179
10	117	53	170
11	79	46	125
12	46	36	82
13	34	15	49
14	31	38	69
15	19	25	44

than the smallest. It appears that Raddatz Rockshelter contains a small background of non-deer faunal material that did not vary widely in abundance through time. To this material was added a widely varying number of deer bones. Figure 4.1 helps make this clear. Deer remains within Raddatz Rockshelter show a truncated normal distribution, suggesting that the mechanism that accumulated them here was distributed in a similar way through time. I note that many accumulation mechanisms may produce a distribution of this sort; slowly increasing and then decreasing use of the site by predators of deer, including people, is one such mechanism. I also note that the movement of materials initially deposited in Levels 5 and 6 to other strata within the site could also produce this pattern, but movement of this sort within Raddatz is not supported by the distribution of non-deer remains. Figure 4.1 also demonstrates that the remains of all vertebrates other than deer have a distribution very different from that of deer, one that approaches uniformity. Given these two very different distributions, it is not surprising that the correlation between rank orders of the number of deer specimens per level and the number of non-deer specimens per level is extremely low ($r_s = .06$). The Raddatz Rockshelter fauna consists of a scanty background of various vertebrate remains that does not change through time, on which a widely varying number of deer remains has been superimposed.

CONCLUSIONS

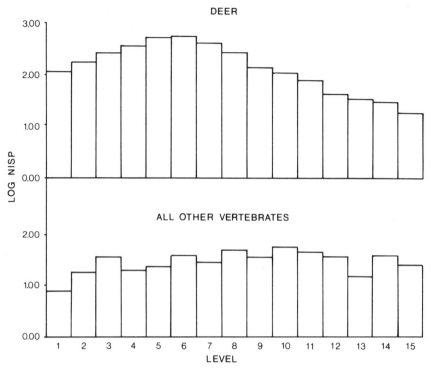

Figure 4.1 The distribution of the remains of deer and all other vertebrates by level within Raddatz Rockshelter.

As a result, it is also no surprise that there is a significant correlation between NISP values per level and the percentage of those specimens that are deer. Each level NISP value is essentially composed of two parts: a variable (deer) and a constant (all other vertebrates). As the number of deer bones increases or decreases per level, so must the fraction of each level NISP that is deer. Any analysis of the Raddatz Rockshelter fauna, then, must take into account the fact that an analysis of the changing relative abundances of deer across levels is primarily an examination of the number of identified deer specimens per level within the site.

Conclusions

Relative taxonomic abundances may be significantly correlated with the size of the samples from which they have been determined. As a result, interpretations of relative abundances may prove to be primarily the interpretation of the size of the samples from which the abundances have been derived. The presence of such a correlation does not necessarily mean that changing sample sizes have caused the changing relative abundances, but it does mean that the cause of the correlation must be explored before further analyses of those abundances are conducted. In most cases that I have examined, the cause of the correlation can readily be attributed to the effects of assemblages with very small numbers of identified specimens (as was the case with the Gatecliff pygmy rabbits and hares, and with the Hidden Cave marmots). In other instances (for instance, Raddatz Rockshelter), the correlation is due to the fact that the relative abundances of a given taxon and the size of the samples from which those relative abundances are drawn are essentially the same thing. In a small number of cases, the cause continues to elude me (see, for example, the composite *Neotoma* samples discussed in Grayson 1984). Importantly, I have found correlations between sample size and relative abundances in about one-third of the published faunal analyses that I have examined (Grayson 1981b).

Fortunately, sample size effects of the sort described in this chapter are extremely easy to detect. While there are many ways to search for them, I have depended here on rank order correlation coefficients because of my belief that, on a single site basis, specimen counts and minimum numbers can usually be shown to provide ordinal scale measures at best (see Chapter 3). The approach I have used for Gatecliff and Hidden Cave can be used generally: rank order the faunal assemblages being examined in terms of sample size (NISP or MNI), then rank them in terms of the relative abundances of the taxon of interest, and then test to see if a significant rank order correlation coefficient emerges. If it does not, analysis can proceed without concern for this particular sample size effect. If it does, then the cause of the correlation must be sought before analysis may proceed.

CHAPTER 5

Taxonomic Richness, Diversity, and Assemblage Size

Cleland (1966) proposed that the continuum of human adaptations be divided into two basic types. Cleland defined a focal subsistence economy as one based on the exploitation of a small number of plant and animal species, and contrasted this adaptation with a diffuse economy, which exploits a wide variety of organisms. Independently, Dunnell (1972) distinguished between intensive and extensive human economic systems, defining them very much as Cleland had defined focal and diffuse systems. Both authors argued that these adaptions were qualitatively different, and the transformation of one into the other would require a "quantum jump" (Cleland 1966:45; Dunnell 1972:79). Both also argued that understanding the prehistoric record of eastern North America would require deep understanding of the evolution of these kinds of subsistence systems, and of the relationship between them. Indeed, general ecological theory specifies that the adaptations of many organisms are, in part, keyed to the distribution and abundance of the organisms on which they prey. Generalists, which feed on a wide variety of organisms in roughly equal numbers, differ in many ways from specialists, which prey on a smaller number of taxa but utilize larger numbers of individuals of those taxa (e.g., Cody 1974; Cody and Diamond 1975; MacArthur 1972).

To address questions of this sort, archaeologists and paleontologists must deal with the number of taxa represented in a given assemblage or in sets of assemblages, and with the distributions of abundance across those taxa. That is, faunal analysts must deal with the taxonomic richness and taxonomic diversity of the faunas with which they work. In this chapter, I discuss some of the difficulties involved in measuring richness and diversity in the archaeological setting.

Taxonomic Richness

In studies of taxonomic richness, analysis is focused on the number of taxa, often species, that have contributed to a faunal assemblage, and on comparing assemblages on the basis of the number of taxa they contain. Taxonomic richness has frequently been addressed in the archaeological literature. Smith (1975a), for instance, utilized species richness as a basic analytic device in his study of the use of vertebrates by Middle Mississippi peoples in the southeastern United States. "The Barker site and the Lilbourn site," Smith (1975a:128) observed, have much higher species counts, minimum numbers of individuals, and poundage estimates" for waterfowl than other sites in his sample, and he related these differences to the location of sites inside and outside of the Mississippi River meander belt zone. Species richness is highly variable across strata within both Gatecliff Shelter and Hidden Cave, and it is tempting to relate this variability to the kinds of accumulation mechanisms that constructed the faunal assemblages of those sites, and to changes in past environments.

Faunal assemblage richness, however, is tightly correlated with the size of the retrieved sample — the number of identified specimens — and this fact must be taken into account whenever richness is analyzed. Paleontologists (see Tipper 1979 and references therein) and ecologists (see May 1975; Peilou 1975; and references therein) are well aware of this fact. Although debates over the

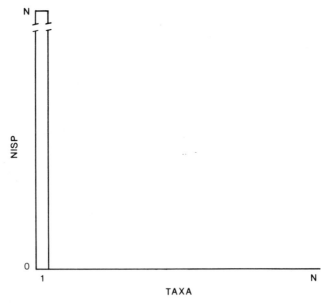

Figure 5.1 Taxonomic frequency structure of a faunal assemblage in which all identified specimens belong to the same taxon.

TAXONOMIC RICHNESS

interpretation of specific sets of richness values arise in these fields (e.g., Raup 1977; Sheehan 1977), the debates themselves show full recognition of the potential hazards of richness indices. Archaeological faunal analysts, on the other hand, seem less aware of the problem, although there are important exceptions (e.g., Styles 1981).

The general nature of the relationship between sample size and taxonomic richness is easy to establish. On the one hand, one can conceive of faunal assemblages in which all identifiable specimens belong to a single taxon; bison kills provide an example (e.g., Wheat 1972). An assemblage of this kind would have the taxonomic frequency structure displayed in Figure 5.1. In sampling a distribution of this sort, only the first specimen identified would add a new taxon, since all additional specimens belong to the same taxon. On the other hand, one can also conceive of a faunal assemblage in which every identified specimen belonged to a different taxon (Figure 5.2). In sampling a distribution

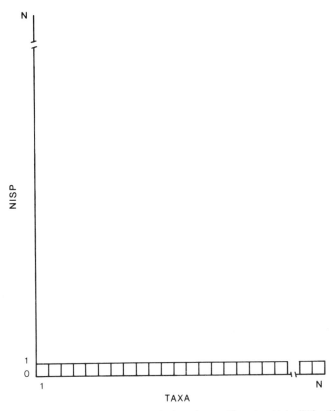

Figure 5.2 Taxonomic frequency structure of a faunal assemblage in which all identified specimens belong to different taxa.

5. TAXONOMIC RICHNESS, DIVERSITY, AND ASSEMBLAGE SIZE

of this sort, each specimen identified would add a new species to the list. The upper and lower boundaries to the relationship between number of taxa encountered and number of specimens identified in faunal assemblages are produced by these two very different kinds of assemblages, and are displayed in Figure 5.3, Lines A and C.

Most archaeological assemblages that I have examined show a very different kind of frequency structure, one in which a few taxa are very abundant, while most are represented by small numbers of specimens or individuals. Several distributions of this sort were illustrated in Chapter 3 (see Figures 3.2 through 3.7); others are illustrated in Grayson (1979b). Yet another, for Hidden Cave Stratum IV, is shown in Figure 5.4. In distributions of this sort, larger and larger samples are needed to detect increasingly rare taxa (see the discussion in Wolff 1975). Sampling a series of distributions of this general shape produces relationships between the number of taxa encountered and the number of specimens identified that are well-described by equations of the form $Y = aX^b$ or $Y = a + b \log X$, in which Y is the number of taxa encountered; X, the number of identified specimens; a, the Y intercept; and b, the slope of relationship between X and Y. In this extremely common situation, richness values among samples of differing sizes cannot be compared in a straightforward fashion.

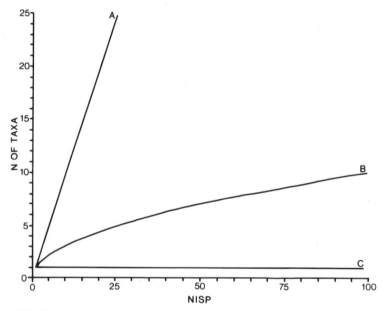

Figure 5.3 The relationship between numbers of identified specimens and numbers of taxa: Lines A and C display the bounds between which this relationship must vary; Line B displays the general form of the usual relationship between these variables.

TAXONOMIC RICHNESS 135

Such a comparison would tell us more about the number of identified specimens in the assemblages involved than it would about the differential richness of those assemblages (see Figure 5.3, Line B).

A simple, and obvious, example of the effects of sample size on species richness may be drawn from the species present within the Hidden Cave Stratum IV mammalian assemblage. The number of identified specimens per

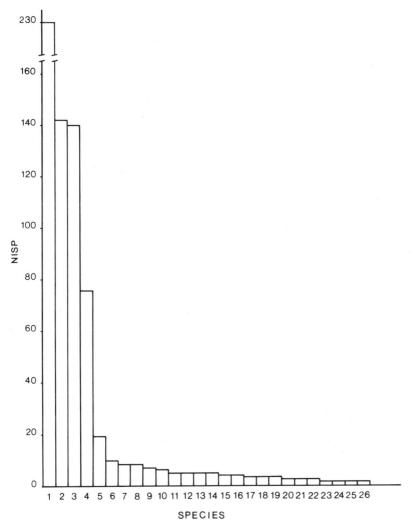

Figure 5.4 The distribution of taxonomic abundances for the Hidden Cave Stratum IV mammals. See Table 5.1 for key to species.

5. TAXONOMIC RICHNESS, DIVERSITY, AND ASSEMBLAGE SIZE

TABLE 5.1

Hidden Cave Stratum IV Species NISP, and a 33% Random Sample of Species NISP[a]

Taxon	Stratum IV NISP	33% Random Sample
Sylvilagus nuttallii (1)	230	90
Neotoma cinerea (2)	142	47
Marmota flaviventris (3)	140	38
Neotoma lepida (4)	76	23
Canis latrans (5)	19	7
Taxidea taxus (6)	10	3
Dipodomys microps (7)	8	4
Peromyscus maniculatus (8)	8	4
Perognathus formosus (9)	7	2
Vulpes vulpes (10)	6	2
Antilocapra americana (11)	5	1
Mustela frenata (12)	5	1
Thomomys bottae (13)	5	1
Spermophilus townsendii (14)	5	0
Microtus montanus (15)	4	1
Ondatra zibethicus (16)	4	0
Perognathus longimembris (17)	3	2
Dipodomys deserti (18)	3	1
Lepus californicus (19)	3	0
Perognathus parvus (20)	2	2
Odocoileus hemionus (21)	2	1
Ammospermophilus leucurus (22)	2	0
Dipodomys ordii (23)	1	1
Canis lupus (24)	1	0
Lynx rufus (25)	1	0
Myotis yumanensis (26)	1	0
	693	231

[a] Numbers in parentheses are used to identify taxa in Figures 5.4 and 5.5.

species in this stratum is shown in Table 5.1 (this list combines specimens securely identified to a given species with those simply referred to that species; thus, the specimens attributed to *Neotoma cinerea* in this table represent the summed NISP values for both *N. cinerea* and *N* cf. *cinerea* in Stratum IV). The distribution of these specimens across taxa is illustrated in Figure 5.4. If the Stratum IV fauna had been sampled less intensively, the number of species represented in that fauna would, of course, have been less. Table 5.1 also presents the results of a 33% random sample (without replacement) of the retrieved Stratum IV mammalian fauna; those results are illustrated in Figure 5.5. Rather than documenting the presence of 26 species from 693 identified

TAXONOMIC RICHNESS

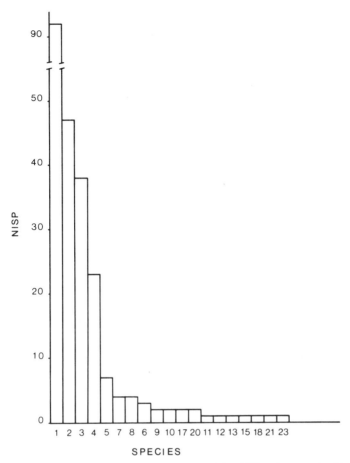

Figure 5.5 The distribution of taxonomic abundances within a 33% random sample of the Hidden Cave Stratum IV mammals specimens identified to the species level. See Table 5.1 for key to species.

specimens, the reduced assemblage documents the presence of 19 species from 231 identified specimens. Of the 10 species above the median NISP value of five, all have survived the sampling process. Of the 12 species beneath this point, only 6 have survived.

This is all very obvious. Certainly, smaller samples will retrieve smaller numbers of species, and certainly if we are truly sampling randomly, the least abundant species will be the last to be detected. The point I wish to make here, however, is that there is an important similarity between sampling the same faunal assemblage twice at different sampling fractions and sampling different

138   5. TAXONOMIC RICHNESS, DIVERSITY, AND ASSEMBLAGE SIZE

faunal assemblages in such a way that varying numbers of identified specimens are retrieved. Both procedures will provide strong correlations between sample size and assemblage richness. Complications enter because the nature of the distributions sampled will differ when different assemblages are involved, but the correlations will be present nonetheless.

Gatecliff Shelter provides an example. The relationship between $\log_{10}$ number of small mammal species per assemblage and $\log_{10}$ NISP per assemblage across all Gatecliff strata that contained at least one specimen identified to the species level is shown in Figure 5.6 (see Table 5.2 for the raw data). The correlation r between these two variables here is .94 ($p < .001$); variation in NISP values can account for 87% of the variance in $\log_{10}$ number of species. The corresponding relationship for the Gatecliff small mammal genera is shown in Figure 5.7 ($r = .91, p < .001$; see Table 5.2 for the raw data). Hidden Cave also

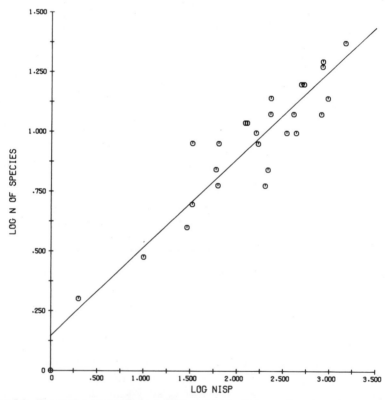

Figure 5.6   The relationship between assemblage richness ($\log_{10}$ number of species per assemblage) and $\log_{10}$ NISP across all Gatecliff Shelter small-mammal assemblages that provided at least one specimen identified to the species level.

TAXONOMIC RICHNESS

TABLE 5.2
Numbers of Specimens Identified to the Species and Genus Levels, Numbers of Species, and Numbers of Genera: Gatecliff Shelter Small Mammals[a]

Stratum	NISP: species	Number of species	NISP: genus	Number of genera
H1	230	14	329	12
H2	827	19	1152	16
H3	1470	24	1935	18
S2	127	11	165	13
H4	835	20	1126	17
H5	522	16	701	13
H6	950	14	1232	13
S6–7	215	7	266	8
S8	59	7	81	8
S9	486	16	636	14
S10	10	3	29	5
S11/12	405	12	546	10
S13	121	11	235	11
S14–16	1	1	1	1
S17	33	5	53	6
S18	2	2	4	3
S19	159	10	291	10
S20	33	9	45	11
S21	—	—	2	2
S22	63	9	94	11
S23	29	4	41	9
S24/25	228	12	440	12
S26–30	62	6	106	8
S31/32	168	9	255	10
S33	432	10	568	12
S37	201	6	273	9
S54	341	10	487	13
S55	—	—	16	2
S56	807	12	1207	11

[a] H, horizon; S, stratum.

provides an example, although there are fewer assemblages with which to work. Figure 5.8 illustrates the relationship between $\log_{10}$ number of mammal species and $\log_{10}$ NISP across the 11 Hidden Cave strata that provided at least one specimen identified to the species level (see Table 5.3 for the raw data). Here, the correlation r between these two variables is .97 ($p < .001$); 93% of the variance in assemblage species richness across the strata of this site can be explained by sample sizes alone. On the generic level, the correlation is equally high: $r = .97$ ($p < .001$; see Figure 5.9 and Table 5.3). Finally, the relationship

140   5. TAXONOMIC RICHNESS, DIVERSITY, AND ASSEMBLAGE SIZE

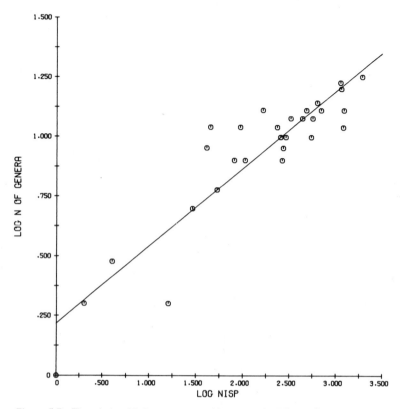

Figure 5.7 The relationship between assemblage generic richness ($\log_{10}$ number of genera per assemblage) and $\log_{10}$ NISP across all Gatecliff Shelter small-mammal assemblages that provided at least one specimen identified to the genus level.

between $\log_{10}$ number of mammal species and $\log_{10}$ NISP for Meadowcroft Rockshelter, southwestern Pennsylvania (Carlisle and Adovasio 1982; Guilday and Parmalee 1982; Guilday et al. n.d.) is shown in Figure 5.10 ($r = .83$, $p < .002$; see Table 5.4). This regression is heavily influenced by Meadowcroft Stratum VIII, which falls more than two standard deviations beneath the predicted $\log_{10}$ number of species value. The Meadowcroft relationship with this outlier removed is shown in Figure 5.11 ($r = .94$, $p < .001$). The regression equations for the Gatecliff, Hidden Cave, and Meadowcroft number of species–NISP relationships are given in Table 5.5.

For the faunas provided by these three sites, analyses of residuals show log transforms of both NISP and numbers of taxa to provide better fits than transforms of NISP alone. For other assemblages, the relationship between these variables is semilogarithmic, as my discussion of Fremont mammal assem-

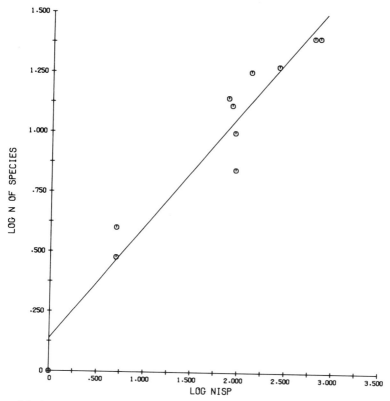

Figure 5.8 The relationship between $\log_{10}$ number of species per assemblage and $\log_{10}$ NISP across all Hidden Cave mammalian assemblages that provided at least one specimen identified to the species level.

blages demonstrates (see below). Finding the most appropriate fit for these relationships is essential if outlying values are to become the focus of attention.

To this point, I have used a series of caves and rockshelters as the basis of my examples, since these sites provide a sequence of stratified assemblages that can be compared within each site. Most comparisons of richness that archaeologists have conducted, however, have been conducted using faunal assemblages from sites scattered across space (e.g., Smith 1975a). It should be evident that such assemblages will also provide taxonomic lists whose lengths will be correlated with the numbers of identified specimens within each assemblage. Faunal assemblages from Fremont sites in Utah show this relationship well.

The Fremont "culture" is an archaeological manifestation known from much of the state of Utah, and from the immediately adjacent parts of Nevada, Idaho,

TABLE 5.3
Numbers of Specimens Identified to the Species and Genus Levels, Numbers of Species, and Numbers of Genera: Hidden Cave Mammals from Unmixed Strata

Stratum	NISP: species	Number of species	NISP: genus	Number of genera
I	75	14	212	13
II	129	18	286	17
III	82	13	189	14
IV	687	25	1374	21
V	597	25	1069	22
VII	253	19	428	18
VIII	1	1	11	1
IX	91	7	158	10
XI	5	3	9	3
XIII	89	10	145	14
XIV	5	4	10	5

and Colorado. Dating to between about A.D. 500 and 1300, Fremont sites document the presence of people subsisting on a combination of hunting, gathering, and horticulture in an area that supported only hunters and gatherers at the time of European contact. The presence of horticulture, as well as specific kinds of ceramics, architecture, and other artifact classes, suggests ties to the Anasazi area of the American Southwest (Jennings 1974, 1978; Madsen 1980a).

TABLE 5.4
Numbers of Specimens Identified to the Species Level and Numbers of Species: Meadowcroft Rockshelter Mammals

Stratum	NISP: species	Number of species
IIA	728	23
IIB	4132	28
III	5630	26
IV	6625	31
V	3284	30
VI	217	15
VII	2033	27
VIII	1110	13
IX	358	15
X	175	15
XI	829	18

TAXONOMIC RICHNESS

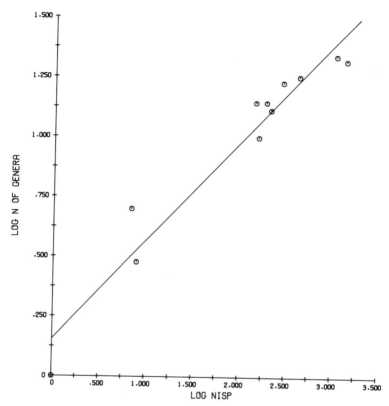

Figure 5.9 The relationship between $\log_{10}$ number of genera per assemblage and $\log_{10}$ NISP across all Hidden Cave mammal assemblages that provided at least one specimen identified to the genus level.

Because the Fremont occupation interrupts what is otherwise an 11,000-year sequence of hunting and gathering adaptations in the arid west, archaeologists have spent much time discussing the rise and fall of this cultural phenomenon.

Since Fremont subsistence practices are at the heart of most attempts to explain Fremont itself, the faunal assemblages provided by these sites have become the target of many studies (e.g., Madsen 1980b, 1982; Parmalee 1980). Some of these studies focus on the number of taxa utilized by Fremont peoples and on the variations in Fremont assemblages in terms of taxonomic richness across space (e.g., Madsen 1980b). It is easy to demonstrate, however, that the richness of Fremont faunal assemblages is highly correlated with the sizes of those assemblages.

The birds from Fremont sites provide an excellent example. Parmalee (1980) has done a real service by summarizing the data available for the avian assem-

5. TAXONOMIC RICHNESS, DIVERSITY, AND ASSEMBLAGE SIZE

TABLE 5.5

Species Richness (R) and Sample Size: Regression Equations and Correlation Coefficients for the Gatecliff, Hidden Cave, and Meadowcroft Mammals

Site	Regression equation	r	p
Gatecliff	$R = 1.40(NISP)^{.37}$	.935	<.001
Hidden Cave	$R = 1.37(NISP)^{.47}$	.965	<.001
Meadowcroft	$R = 4.62(NISP)^{.21}$	.834	<.002

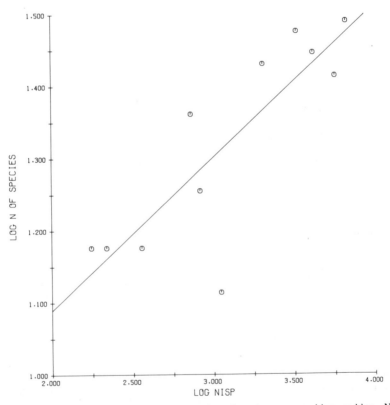

Figure 5.10 The relationship between $\log_{10}$ number of species per assemblage and $\log_{10}$ NISP across all Meadowcroft Rockshelter mammalian assemblages.

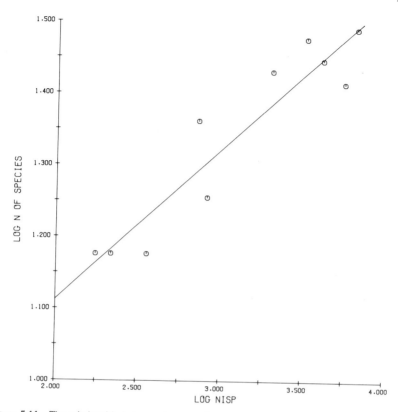

Figure 5.11 The relationship between $\log_{10}$ number of species per assemblage and $\log_{10}$ NISP across all Meadowcroft Rockshelter mammalian assemblages except Stratum VIII.

blages from 11 Fremont open sites. Data on assemblage size and taxonomic richness for those faunas are presented in Table 5.6, while Figure 5.12 displays the relationship between $\log_{10}$ number of species and $\log_{10}$ NISP for those 11 assemblages ($r = .92$, $p < .001$). Clearly, any analysis of avian richness among Fremont sites must take into account the role of sample size in determining that richness, or else the analyst runs the risk of interpreting as differences in richness what are actually differences in sample size per assemblage.

The mammals from Fremont sites show similar correlations. Table 5.7 presents the raw data on the number of specimens identified to the species level, and the number of species identified from those specimens, from 17 Fremont open sites. I note that there are many more than 17 such sites, but a large number were excavated prior to the mid-1960s, when faunal remains were often inadequately reported or even neglected. Further, I have included only open sites in my analysis, thus eliminating such sites as the extremely

TABLE 5.6

Numbers of Specimens Identified to the Species Level and Numbers of Species for 11 Fremont Avian Assemblages

Site	NISP: species	Number of species
Levee	582	34
Bear River 3	353	27
Pharo Village	141	16
Injun Creek	137	26
Bear River 1	130	30
Knoll	108	16
Nephi	48	14
Warren	44	15
Evans Mound	41	9
42SL19	15	9
Black Rock III	9	5

TABLE 5.7

Numbers of Specimens Identified to the Species Level and Numbers of Species for 17 Fremont Mammalian Assemblages

Site	NISP: species	Number of species	Reference
Evans Mound	4601	16	Berry 1972
Median Village	2782	7	Marwitt 1970
Bear River 1	1859	13	Aikens 1966
Injun Creek	1859	12	Aikens 1966
Bear River 2	943	14	Aikens 1967
Bear River 3	827	13	Shields and Dalley 1978
Nephi Mound I	781	8	Sharrock and Marwitt 1967
Pharo Village	755	17	Marwitt 1968
Snake Rock	737	12	Aikens 1967
Caldwell Village	307	10	Ambler 1966
Backhoe Village	100	7	Madsen and Lindsay 1977
Old Woman	82	10	Taylor 1957
Windy Ridge	37	5	Madsen 1975
Felter Hill	16	1	Shields 1967
Whiterocks	16	3	Shields 1967
Innocents Ridge	9	3	Schroedl and Hogan 1975
Poplar Knob	2	2	Taylor 1957

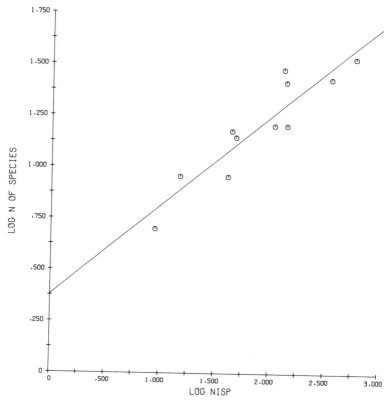

Figure 5.12 The relationship between $\log_{10}$ number of species and $\log_{10}$ NISP for 11 Fremont avian assemblages.

well-reported Hogup Cave (Aikens 1970). I have not included caves and rockshelters because of the extreme difficulty of knowing the mechanisms that accumulated the faunas within these sites. Although such problems are by no means solved by the use of open sites alone, they are certainly lessened, since the mechanisms that deposit faunal material in open archaeological sites would appear to be less varied than those that deposit them in rockshelters and caves.

The relationship between $\log_{10}$ number of mammal species and $\log_{10}$ NISP for the 17 Fremont mammalian assemblages is shown in Figure 5.13. The residuals for this relationship exhibit a strong systematic trend when plotted against either NISP or against predicted number of species (Y) values: assemblages with high NISP or high predicted Y values tend to fall beneath the regression line, while those with low NISP or low predicted Y values tend to fall above this line (the runs test probability for this pattern is $<.10$; see Figure 5.14). Figure 5.15 displays the relationship between untransformed numbers of

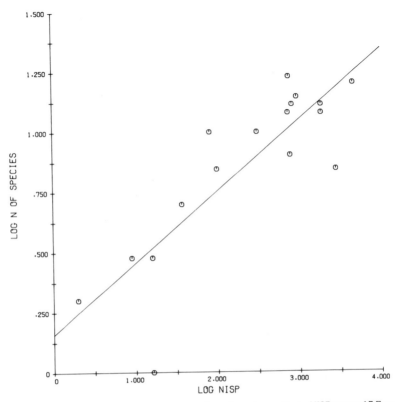

Figure 5.13 The relationship between $\log_{10}$ number of species and $\log_{10}$ NISP across 17 Fremont mammalian assemblages.

species and $\log_{10}$ NISP. Plots of residuals against NISP and against predicted Y values exhibit no systematic trend (the runs test probability for this pattern is > .50; see Figure 5.16). For these 17 assemblages, then, the relationship between species richness and the number of identified specimens is semilogarithmic in form. That the nature of this relationship must be taken into account in discussing the species richness of Fremont mammalian assemblages should be clear. Let us say, for instance, that an analyst wanted to compare the species richness of mammalian faunas from marsh-side Fremont sites to sites located in non-marsh settings. It might be tempting to compare the 13 mammal species present in the fauna from the marsh-side site of Bear River 1 to the seven species provided by Backhoe Village, and to conclude that the apparent differences in richness reflect environmental setting. But the differences are only apparent. Bear River 1 has 1859 specimens identified to the species level, while Backhoe Village has only 100 specimens identified to that level. The regression equation

TAXONOMIC RICHNESS

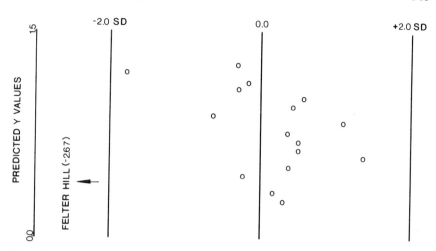

Figure 5.14 The relationship between $\log_{10}$ number of species (Y) and $\log_{10}$ NISP across 17 Fremont mammalian assemblages: plot of residuals in unit normal deviate form against predicted Y values.

for the 17 Fremont site (Table 5.8) predicts that Bear River 1 should have 12.9 species of mammals, while Backhoe Village should have 7.6 species. In fact, then, the number of species at both sites is very close to predicted values, and comparisons of these values tell us more about sample sizes than about true richness. The primary information on differences in species richness among these assemblages that emerges from my analysis involves two sites, both of which have richness values that fall more than two standard deviations from predicted values. Pharo Village has a far greater number of species than would be predicted from the general relationship between numbers of specimens and numbers of species (17 observed species as opposed to 11.4 predicted), while Median Village has many fewer species than predicted (7 observed compared to 13.8 predicted). Any interpretation of richness across Fremont assemblages must thus focus on deviations from predicted values, that is, on the residuals, and not on the simple number of species per faunal assemblage. Note also that if the log-log relationship had been mistakenly used as the basis for the analysis, neither Pharo Village nor Median Village would have been detected as outliers, and we would have instead focused on the Felter Hill site as being the only outlier in this set of assemblages.

The interpretation of this particular set of Fremont richness values might or might not focus on the subsistence meaning of the Pharo Village and Median Village figures. Since the independent variable in the regression is the number of identified specimens, the positions of the Pharo and Median village sites may relate to mammalian procurement practices, but they may also relate to butch-

TABLE 5.8
Assemblage Species Richness (R) and Sample Size: Regression Equations and Correlation Coefficients for 11 Avian and 17 Mammalian Faunal Assemblages for Fremont Sites

	Regression equation	r	p
Avian Assemblages	$R = 2.57(NISP)^{.42}$	.916	<.001
Mammalian Assemblages	$R = -0.83 + 4.18 \log_{10} NISP$	.831	<.001

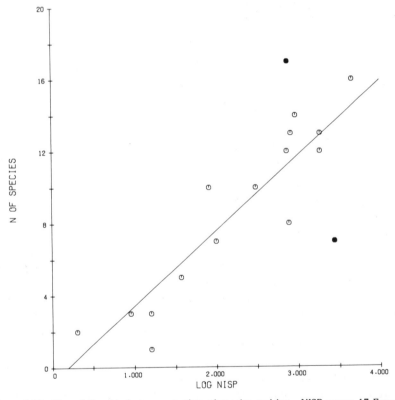

Figure 5.15 The relationship between number of species and $\log_{10}$ NISP across 17 Fremont mammalian assemblages. Pharo Village (above line) and Median Village (below line) are indicated by closed circles.

RAREFACTION AND SPECIES-ABUNDANCE DISTRIBUTIONS

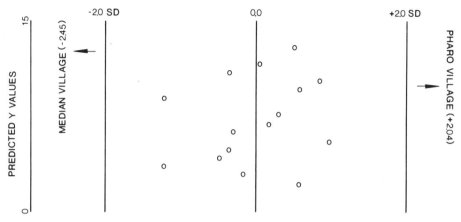

Figure 5.16 The relationship between number of species (Y) and $\log_{10}$ NISP across 17 Fremont mammalian assemblages: plot of residuals in unit normal deviate form against predicted Y values.

ering. It is not possible to extract information on butchering in particular, and on bone fragmentation in general, from Fremont site reports, since the published analyses were not conducted with this problem in mind. As a result, it is not possible to address this taphonomic issue here. It is also not possible to examine the placement of Pharo and Median villages on plots that display the relationship between MNI and numbers of species for Fremont assemblages because MNI values have generally not been provided in Fremont reports. As I have discussed, however, it is likely that such an examination would also direct us toward currently unavailable information on bone fragmentation.

Rarefaction and the Comparison of Species-Abundance Distributions

Because simple counts of the number of species per faunal sample are affected by sample size, a technique developed by Sanders (1968) has become somewhat popular among ecologists and paleontologists. Sanders (1968) wished to compare the species richness values of a number of samples of marine invertebrates taken from various regions and with widely varying sample sizes. In order to avoid sample size effects, Sanders developed a method that allowed him to estimate the number of species a given assemblage would have at smaller sample sizes. The question Sanders asked was, on the surface, quite simple: How many species would a given sample have were that sample smaller, or "rarer," in individuals? The method he developed is, thus, called *rarefaction*.

By calculating estimated species numbers for a series of sample sizes smaller than the total recovered sample, Sanders was able to generate curves that represented the estimated increase in number of species per sample as the number of individuals in that sample increased (Sanders' method of calculation was simple, and is clearly explained in Sanders 1968). He generated such curves for each of his marine invertebrate samples, and then compared richness values by qualitatively comparing those curves. His method of calculating estimated species values for reduced sample sizes was later shown to be flawed (Hurlbert 1971), and his original approach has now been replaced by a more accurate one (see the excellent review by Tipper 1979).

Sanders' method was developed for situations in which each specimen in the paleontological or modern sample represents an individual. The refinements in his approach were also developed under the assumption that the units being counted are true individuals. Accordingly, subsequent paleontological applications of the method have all been to communities of marine invertebrates, in which the censusing unit represents a true individual. Unfortunately, since nothing comparable to this unit exists for archaeological vertebrate faunas, rarefaction cannot be appropriately applied to these faunas. For this reason, I am very wary of the results obtained by Styles (1981) in her application of Sanders' original rarefaction methodology to archaeological vertebrates using minimum numbers of individuals as the censusing unit (nonetheless, Styles' recognition of the problems involved in interpreting species richness in archaeological faunas is of high importance, and it is difficult to understand why her work has not been more widely discussed than it has been).

Thus, rarefaction would not seem to provide an acceptable method for extracting validly comparable richness values for archaeological faunal assemblages of different sizes that cannot be censused using true individuals. However, as I noted at the beginning of this chapter, archaeologists are often interested not simply in the number of species present in given sets of assemblages, but in both the numbers of species present and in the distributions of abundance across those species. That is, archaeologists are often more interested in taxonomic diversity than they are in taxonomic richness (for discussions of these concepts, see May 1975; Peet 1974; and Pielou 1975). Although a number of specific indices have been developed in ecology to measure diversity (see the next section), paleontologists wary of such indices might turn instead to the comparison of curves generated by the rarefaction method. Tipper (1979), however, has observed that rather than using either diversity indices or rarefaction curves, species-abundance distributions can be directly compared using the two-sided Smirnov test in order to determine whether or not they are of the same form. This approach can be applied to archaeological faunal assemblages using either minimum numbers or numbers of identified

RAREFACTION AND SPECIES-ABUNDANCE DISTRIBUTIONS 153

specimens, as long as these counting units have achieved an ordinal measure of taxonomic abundance.

An example of this approach can be drawn from the Fremont vertebrate assemblages. Since these assemblages will be the empirical focus of much of the rest of this chapter, further background on the debate concerning Fremont subsistence practices will be helpful here. Madsen (1980b, 1982) has argued that in certain parts of Fremont territory, and in particular in the eastern Great Basin, marsh resources were extremely important in the development of what appear to have been sedentary Fremont villages. In particular, he has argued that in marsh-side settings, floral and faunal resources may have been sufficiently abundant and predictable to allow the development of sedentary villages, with domestic crops acting as supplemental resources. In settings in which marsh resources were not available, he argues, domestic crops were crucial in allowing the development of sedentary villages, with nondomestic food sources secondary. If Madsen is correct, then analysis of the structure of Fremont vertebrate faunal assemblages should show significant differences between those sites located adjacent to marshes and those located in other settings.

It is easy to show that even adjacent Fremont sites are characterized by faunal assemblages that are significantly different in terms of the representation of particular species of mammals. The Bear River 1 site, for instance, is located on the south side of the Bear River, northwestern Utah, at an elevation of 1285 m and adjacent to alluvial marshland (Aikens 1966). At the same elevation and 2.4 km to the southwest is the Bear River 3 site, which is also adjacent to the Bear River and to alluvial marshland (Shields and Dalley 1978). Bear River 1 has been radiocarbon dated to 885 ± 120 A.D. (Aikens 1966). Bear River 3 has provided a single date of 500 ± 110 A.D., but Shields and Dalley (1978) note that this date is out of line with ceramic information from the site. The ceramics suggest that Bear River 3 falls close in time to yet another Fremont site in this complex, Bear River 2, dated to 955 ± 105 A.D. (Aikens 1967). An independent seriation based on projectile points suggests that Bear River 1 is somewhat, but not much, younger than Bear River 3 (Holmer and Weder 1980). Thus, Bear River Sites 1 and 3 are very similar to one another in their environmental setting, and seem to fall very close to one another in time.

They are not, however, very close to one another in the abundances of the species that compose their mammalian faunas. Table 5.9 presents the number of identified specimens per mammal species at these two sites, with the value of four rare species combined in order to ensure that all expected values for chi-square analysis are greater than one. Chi-square analysis shows that these assemblages are not homogeneous as regards the abundances of their component taxa ($\chi^2 = 280.39$, $p < .001$). Calculation of adjusted residuals (Everitt

TABLE 5.9

Numbers of Identified Specimens (NISP) per Mammalian Species at the Bear River 1 (BR1) and Bear River 3 (BR3) Sites[a,b]

Species	NISP, BR1	NISP, BR3
Bison bison	1541	670
Cervus elaphus	173	0
Odocoileus hemionus	87	26
Ondatra zibethicus	28	60
Erethizon dorsatum	8	1
Marmota flaviventris	8	0
Ovis canadensis	5	0
Mustela vison	3	29
Castor canadensis	2	2
Mephitis mephitis	1	9
Lepus californicus	0	9
Antilocapra americana	0	10
Lutra canadensis	0	7
*Four rare species	3	4
	1859	827
*Microtus montanus	1	1
Canis latrans	1	0
Ursus americanus	0	2
Lynx rufus	1	1

[a] From Aikens (1966) and Shields and Dalley (1978).
[b] $\chi^2 = 280.39$, $p < .001$.

1977; adjusted residuals are read as standard normal deviates) provides details on the precise source of this highly significant χ^2 value (Table 5.10). Compared to Bear River 3, elk specimens are greatly overrepresented at Bear River 1, while specimens of antelope, striped skunk, mink, black-tailed jackrabbit, muskrat, and river otter are greatly underrepresented. This comparison does not necessarily tell us about differences in faunal capture, of course, since the significant differences in numbers of identified specimens could have come about through such things as differences in butchering practices. This analysis does, however, direct us to seek further information on those taxa whose NISP-based abundances are causing the significant differences.

The chi-square analysis just presented may indicate a significant difference in the capture and utilization of mammalian resources by the occupants of Bear River Sites 1 and 3. However, direct comparison of the species-abundance distributions provided by these two sites yields a very different sort of information. The two-sided Smirnov test compares the cumulative distribution func-

TABLE 5.10
Adjusted Residuals for the Bear River 1 (BR1) and Bear River 3 (BR3) Mammalian Species Assemblages

Species	Adjusted residuals	
	BR1	BR3
Bison bison	1.21	−1.18
Cervus elaphus	9.04[a]	−9.05[a]
Odocoileus hemionus	1.84	−1.83
Ondatra zibethicus	−7.70[a]	7.72[a]
Erethizon dorsatum	1.28	−1.28
Marmota flaviventris	1.89	−1.89
Ovis canadensis	1.49	−1.49
Mustela vison	−7.43[a]	7.40[a]
Castor canadensis	−0.83	0.83
Mephitis mephitis	−4.04[a]	4.06[a]
Lepus californicus	−4.49[a]	4.50[a]
Antilocapra americana	−4.72[a]	4.75[a]
Lutra canadensis	−3.95[b]	3.97[b]
Four rare species[c]	−1.60	1.61

[a] $p < .001$.
[b] $p < .01$.
[c] See Table 5.9.

tions of two samples, and provides a test of the null hypothesis that those functions do not differ significantly (Conover 1971). Unlike the chi-square test for homogeneity that I have presented for the Bear River Sites 1 and 3 faunal assemblages, the two-sided Smirnov test compares cumulative distribution functions and does not take into account the actual species involved in those distributions. As will become clear, two samples can differ greatly in the relative abundances of the taxa that they share, but not differ significantly in their cumulative distribution functions. Such a result suggests that there may be some underlying structure in the relative proportions of species represented in a pair of faunal assemblages that is independent of the specific species involved.

To conduct a two-sided Smirnov test using numbers of identified specimens per species in an archaeological assemblage, the analyst simply tabulates, in ascending order, the number of species having a given NISP value for each assemblage, and then converts that information into cumulative distribution functions. The test statistic T' is the absolute value of the maximum difference between these functions; either tables or, for large samples, direct calculations, can then be used to determine the corresponding probability values (Conover 1971:309–314).

5. TAXONOMIC RICHNESS, DIVERSITY, AND ASSEMBLAGE SIZE

TABLE 5.11

Calculation of the Smirnov Test Statistic for the Bear River 1 and Bear River 3 Mammalian Species Assemblages[a,b]

NISP	M_i	V_i	m_i	v_i	v^*_i
1	4	.308	3	.231	.077
2	1	.385	2	.385	.000
3	1	.462	—	.385	.077
5	1	.538	—	.385	.153
7	—	.538	1	.462	.076
8	2	.692	—	.462	.230
9	—	.692	2	.615	.077
10	—	.692	1	.692	.000
26	—	.692	1	.769	−.077
28	1	.769	—	.769	.000
29	—	.769	1	.846	−.077
60	—	.769	1	.923	−.154
87	1	.846	—	.923	−.077
173	1	.923	—	.923	.000
670	—	.923	1	1.000	−.077
1541	1	1.000	—	1.000	.000

[a] See text for explanation of column entries.
[b] $T' = |v^*_i|\max = .230 (p > .20)$.

Table 5.11 provides such a calculation for the Bear River 1 and 3 mammalian faunal assemblages. The first column (NISP) provides the number of identified specimens per species for all species at both sites. The second column (M_i; symbols follow the usage of Tipper 1979) provides the number of species represented at the given NISP value at Bear River 1, while the third column (V_i) provides the cumulative distribution function derived from the data in column two. The next two columns (m_i and v_i) provide comparable information for Bear River 3; the last column (v^*_i) gives the magnitude of difference between the cumulative distribution functions at each value of NISP. In this example, the test statistic T', the maximum absolute difference between the two functions, is .230 ($p > .20$). Thus, these two cumulative distribution functions are not significantly different: they could have been drawn from the same population. While the chi-square analysis demonstrates highly significant differences in the relative abundances of specific species at these sites, the Smirnov test suggests that there is nonetheless an underlying similarity in the structure of the species-abundance distributions of these two faunal assemblages, one that is independent of the species involved (see the discussion in Pielou 1975).

Comparisons of the species-abundance distributions of other Fremont sites

TABLE 5.12

Smirnov Test Statistics T' for a Series of Fremont Mammalian Species Assemblages

Comparison	T'	p
Bear River 1 – Bear River 2	.165	> .20
Bear River 1 – Bear River 3	.230	> .20
Bear River 2 – Bear River 3	.171	> .20
Bear River 2 – Pharo Village	.315	> .20
Bear River 2 – Median Village	.428	.20 > p > .10
Pharo Village – Median Village	.286	> .20
Bear River 1 – Backhoe Village	.406	> .20
Backhoe Village – Median Village	.286	> .20

also suggests an underlying similarity in the structure of these distributions, whether or not the sites are located next to marshes. Table 5.12 presents the results of eight such comparisons, none of which are significant. Bear River 1, 2, and 3 are located next to one another in the setting described above; Pharo Village is in a streamside setting in central Utah at an elevation of 1830 m, and dates to approximately 1200 A.D. (Marwitt 1968); Median Village is in southeastern Utah at an elevation of 1783 m, also in a streamside setting, and dates to approximately 950 A.D. (Marwitt 1979); finally, Backhoe Village is at an elevation of 3045 m in the Sevier River valley of central Utah, 4.8 km west of the modern course of the river, and dates to approximately 850 A.D. (Madsen and Lindsay 1977).

The fact that numbers of identified specimens (or MNI) must be used in applications of the two-sided Smirnov test to archaeological vertebrate faunas introduces obvious interpretive difficulties. In addition, there are difficulties with the use of this test in the ecological and paleoecological setting even when true individuals are being counted; these problems have been well-discussed by Conover (1971), Pielou (1975), and Tipper (1979). Because the test assumes the variables to be continuous, the test is conservative when applied to discrete variables, increasing the chances of concluding that two cumulative distribution functions are statistically identical when, in fact, they are not. In addition, two samples of very different size can have very different species-abundance distributions even though drawn from the same population, increasing the chances of concluding that cumulative distribution functions are significantly different when, in fact, they are not. The Smirnov test also assumes that the samples involved are random (very unlikely in the case of the Fremont assemblages discussed here), and that at least an ordinal measurement scale has been achieved (see Chapter 3). In addition, when the number of species involved in the comparison is small, as it is in all the Fremont examples I have presented, the test is not powerful, but this difficulty is not confined to the two-sided Smirnov test.

Of the 136 different species-abundance distribution comparisons possible for the 17 Fremont mammalian assemblages, I have presented only eight. In addition, I have not examined the perhaps crucial avian species-abundance distributions. My purpose here is not to present a full analysis of Fremont vertebrate assemblages, but instead to discuss some ways in which the hypotheses that have been presented concerning Fremont subsistence could be tested quantitatively using currently available information from those assemblages. A combination of chi-square analysis, including analysis of residuals, and comparisons of species-abundance distributions may provide a powerful way to approach such questions. Whether all Fremont vertebrate faunas possess statistically identical species-abundance distributions is unknown to me but if they do, it is possible that such similarities relate to an underlying structural similarity in the procurement or processing of animals by Fremont peoples, even though the precise species involved, and the relative abundances of those species, differ significantly from site to site. While I have used Fremont as an example, the approach I have discussed can be used in many settings. Even if an analysis of this sort results primarily in the more precise definition of taphonomic questions, we will have increased our knowledge of what it is we need to know about archaeological faunal assemblages to anwer the questions we ask.

Diversity Indices and Sample Size

I have mentioned that ecologists have constructed many indices to measure taxonomic diversity, indices that consider both the number of taxa in a sample and the distribution of individuals across those taxa (see the reviews in May 1975; Peet 1974; and Pielou 1975). Archaeologically, these measures can be easily applied since the information required — identification of the taxa present in an assemblage, and a measure of the abundances of those taxa — is readily extracted from faunal data and is routinely supplied in published faunal reports.

It is important to realize that some archaeological applications of diversity measures are prone to sample size effects even when the diversity measure itself is not necessarily a function of the number of items counted (Grayson 1981b). Such effects frequently result from underlying relationships between NISP and relative taxonomic abundance (see Chapter 4).

An example of the often-subtle nature of these effects can be drawn from an examination of MNI-based diversity indices. As I have discussed (Chapter 2), the minimum number of individuals per taxon is a function of the number of identified specimens per taxon; the equation that relates these two variables is generally of the form $MNI = a(NISP)^b$. Since this is the case, whenever the

DIVERSITY INDICES AND SAMPLE SIZE

relative abundances of taxa as measured by NISP are a function of sample size (Chapter 4), diversity measures based on MNI will also be a function of sample size.

Take, for instance, the Shannon index, $-\Sigma p_i \ln p_i$, in which p_i is simply the proportion of individuals in the sample that falls in species i. This measure of diversity has frequently been used in ecological studies, and has also been used in archaeological analysis (Wing 1963, 1975; see MacArthur 1972 and Pielou 1975 for discussions of this index). Using minimum numbers as the measure of species abundance, the value p_i in the Shannon index is calculated as $MNI_i(100)/\Sigma MNI$. Thus, the index may be rewritten as

$$-\Sigma \frac{MNI_i(100)}{\Sigma MNI} \ln \frac{MNI_i(100)}{\Sigma MNI}$$

Since $MNI = a(NISP)^b$, this index also equals

$$-\Sigma \frac{a(NISP)^b{}_i(100)}{\Sigma a(NISP)^b} \ln \frac{a(NISP)^b{}_i(100)}{\Sigma a(NISP)^b}$$

Clearly, if the value $NISP_i/\Sigma NISP$ is a function of sample size, then the calculated diversity indices will also vary as a function of sample size.

The diversity indices presented by Wing (1963) illustrate this effect well. Wing (1963) analyzed the vertebrate remains from the Jungerman site, located near the east coast of Florida. As part of this analysis, she calculated Shannon diversity indices for the fauna from 9 of the 13 excavated levels of this site, 4 of which pertained to the earlier St. John's phase, the remaining 5 pertaining to the St. John's II phase. She discovered that the diversity measures for St. John's II levels were less than those for the St. John's I levels, and suggested several cultural explanations to account for this apparent decrease in diversity through time.

Since the archaeologically applied Shannon index will vary as a function of sample size if the values $NISP_i/\Sigma NISP$ vary as a function of sample size, it is essential to examine this relationship in the Jungerman data. Table 5.13 presents those values for the two most abundant taxa at Jungerman: gopher tortoise *(Gopherus polyphemus)* and sharks (Spualiformes). Spearman's r_s between $NISP_i/\Sigma NISP$ and total level NISP for gopher tortoise is $-.87$ ($p < .01$); that between $NISP_i/\Sigma NISP$ and total level NISP for shark is $+.95$ ($p < .01$). Clearly, it is reasonable to supppose that these values are being determined by sample size: as sample size per level increases, the relative abundance of gopher tortoise decreases, while that for shark increases (since these are percentage values, the two sets of percentages are not, of course, independent of one another: see Chapter 2).

Because the Jungerman $NISP_i/\Sigma NISP$ values vary with sample size, Wing's diversity measure ($-\Sigma p_i \ln p_i$) should also vary with sample size. Table 5.14 presents minimum numbers and specimen counts by level at the Jungerman

5. TAXONOMIC RICHNESS, DIVERSITY, AND ASSEMBLAGE SIZE

TABLE 5.13
NISP$_i$/Σ NISP for Gopher Tortoise and Shark and Total Numbers of Identified Specimens per Level (Σ NISP), Jungerman Fauna[a]

	Gopher tortoise		Shark			
Level	NISP$_i$(100)/Σ NISP	Rank	NISP$_i$(100)/Σ NISP	Rank	Σ NISP	Rank
2	82	2	00	8.5	17	8
3	45	3	00	8.5	11	9
4	89	1	02	7	45	6
5	35	4	06	6	34	7
8	07	9	61	1	180	2
9	13(13.2)	8	45	2	220	1
10	13(13.4)	7	37	3	134	3
11	19	5	28	4	54	5
12	17	6	25	5	63	4

[a] Spearman's r_s, NISP$_i$/Σ NISP gopher tortoise − Σ NISP = −0.87 ($p < .01$), Spearman's r_s, NISP$_i$/Σ NISP shark − Σ NISP = +0.95 ($p < .01$).

site, as well as the diversity values calculated for those levels. The relationship predicted on the basis of the behavior of NISP$_i$/Σ NISP values does, in fact, occur: Spearman's r_s between MNI and diversity is +.87 ($p < .01$); that between NISP and diversity is +.85 ($p < .01$).

The general conclusion is clear. If the values NISP$_i$/Σ NISP vary with sample size, diversity indices based on those values will also so vary. As a result, the meaning of such indices becomes clouded: it may not be at all clear whether they are measuring the diversity of an archaeological fauna, or the size of the faunal samples per stratum or per level retrieved from the site in question.

Although Wing's application of the Shannon index to the Jungerman data is thus weakened, this does not mean that diversity indices cannot be usefully applied to archaeological faunal data. Care must be taken to ensure that sample size is not being measured, and detailed taphonomic information may be needed to interpret the results validly, but even if the indices simply direct the analyst to the kinds of taphonomic information needed to arrive at a valid interpretation, they will have been of value.

The Fremont vertebrates again provide an example. Rather than using the Shannon index to measure Fremont bird and mammal assemblage diversities, I have used the reciprocal of Simpson's index, $1/\Sigma\ p_i^2$, where p_i again represents the proportion of the individuals in the total collection that fall in species i. As MacArthur (1972:189) noted, this index represents the "number of equally common species" in a given sample of species; the higher the value, the more evenly distributed the individuals across species.

TABLE 5.14
Diversity Indices and Sample Sizes, Jungerman Fauna[a,b]

Level	MNI	Rank	NISP	Rank	Diversity	Rank
2	6	9	17	8	1.25	8
3	7	7.5	11	9	1.27	6.5
4	7	7.5	45	6	1.27	6.5
5	17	6	34	7	1.17	9
8	38	2	180	2	2.66	3
9	33	3	220	1	2.70	2
10	40	1	134	3	2.98	1
11	20	5	54	5	2.32	4
12	22	4	63	4	2.29	5

[a] From Wing (1963); all diversity indices have been recalculated from data presented by Wing.
[b] Spearman's r_s, MNI-diversity = $+0.87 (p < .01)$, Spearman's r_s, MNI-diversity = $+0.85 (p < .01)$.

For the Fremont mammal assemblages, I calculated $1/\Sigma\, p_i^2$ values using numbers of identified specimens per taxon, eliminating from consideration those assemblages with 30 identified specimens or less (Innocents Ridge, Poplar Knob, Whiterocks, and Felter Hill). I did not calculate MNI-based diversity values for the mammalian assemblages, since minimum numbers are generally not provided in the Fremont literature. In calculating avian species diversities, I also eliminated assemblages with less than 30 identified specimens (Black Rock III and 42SL19). However, because Parmalee (1980) provides both specimen counts and minimum numbers for all of his assemblages, I have been able to calculate both NISP-based and MNI-based $1/\Sigma\, p_i^2$ values for those assemblages. The results are provided in Tables 5.15 (mammalian assemblages), 5.16 (avian assemblages, based on NISP), and 5.17 (avian assemblages, based on MNI).

Recall that Madsen's hypothesis concerning Fremont settlement patterns specifies that the high diversity, abundance, and predictability of marsh resources played an important role in the development of sedentary Fremont villages in marsh-side settings, and that the lack of such resources led to a secondary role for nondomesticated food resources in other settings. If his hypothesis is correct, then the diversity values I have calculated should differ in predictable ways between sites that are, and are not, located adjacent to marshes. Before exploring that issue, however, it is essential to establish that those values are not merely reflecting sample size.

In only one case is that fact easy to establish. Spearman's r_s between $1/\Sigma\, p_i^2$ and NISP for the nine Fremont avian assemblages in my analysis is $+.16$:

TABLE 5.15
Diversity Values ($1/\Sigma\, p_i^2$) and Total Numbers of Identified Specimens (Σ NISP) for 13 Fremont Mammalian Assemblages

Site	$1/\Sigma\, p_i^2$	Σ NISP
Backhoe Village	8.020	100
Bear River 1	1.432	1859
Bear River 2	1.418	943
Bear River 3	1.505	827
Caldwell Village	3.486	307
Evans Mound	3.516	4601
Injun Creek	1.176	1859
Median Village	1.373	2782
Nephi Mound 1	1.473	781
Old Woman	5.669	82
Pharo Village	3.662	755
Snake Rock	1.560	737
Windy Ridge	2.559	37

there is no significant relationship between calculated diversities and numbers of identified specimens in this data set. For the Fremont mammals and the Fremont MNI-based avian diversities, however, this is not the case.

Spearman's r_s between $1/\Sigma\, p_i^2$ and NISP for the Fremont mammals is $-.62$ ($p < .05$; see Table 5.15). There is a strong tendency for assemblages with larger numbers of identified specimens to be less diverse than assemblages with fewer specimens. It is easy to show how this has happened, but more difficult to explain the phenomenon with the information at hand.

TABLE 5.16
Diversity Values ($1/\Sigma\, p_i^2$) and Total Numbers of Identified Specimens (Σ NISP) for Nine Fremont Avian Species Assemblages

Site	$1/\Sigma\, p_i^2$	Σ NISP
Bear River 1	14.519	130
Bear River 3	5.275	353
Evans Mound	5.548	41
Injun Creek	3.352	130
Knoll	5.475	107
Levee	10.932	582
Nephi Mound 1	5.408	48
Pharo Village	3.112	141
Warren	6.722	44

DIVERSITY INDICES AND SAMPLE SIZE

TABLE 5.17
Diversity Values ($1/\Sigma\, p_i^2$) and Total Minimum Numbers of Individuals (Σ MNI) for Nine Fremont Avian Species Assemblages

Site	$1/\Sigma\, p_i^2$	Σ MNI
Bear River 1	13.056	63
Bear River 3	14.187	74
Evans Mound	8.000	12
Injun Creek	10.637	44
Knoll	10.286	36
Levee	17.906	139
Nephi Mound 1	10.526	20
Pharo Village	7.934	38
Warren	12.522	24

Using numbers of identified specimens as the abundance measure, $1/\Sigma\, p_i^2$ is calculated as the reciprocal of the sum of ($NISP_i/\Sigma$ NISP)^2. As was the case with the MNI-based diversity measures used by Wing (1963), if $NISP_i/\Sigma$ NISP varies with sample size, so will the diversity measure. Among Fremont avian assemblages, there is no significant correlation between $NISP_i/\Sigma$ NISP and Σ NISP, where i represents the most abundant avian species in the collection ($r_s = +.20$; see Table 5.18). Among the Fremont mammalian assemblages, however, a significant correlation is present. Spearman's r_s between $NISP_i/\Sigma$ NISP and Σ NISP, where i now represents the most abundant mammalian species in the collection, is $+.67$ ($p < .025$; see Table 5.19). It would appear

TABLE 5.18
Numbers of Identified Specimens for the Most Abundant Taxon ($NISP_i$), Total Numbers of Identified Specimens (Σ NISP), and $NISP_i/\Sigma$ NISP for Nine Fremont Avian Species Assemblages

Site	$NISP_i$	Σ NISP	$NISP_i/\Sigma$ NISP
Bear River 1	18	130	.138
Bear River 3	133	353	.377
Evans Mound	11	41	.268
Injun Creek	69	130	.531
Knoll	38	107	.355
Levee	114	582	.196
Nephi Mound 1	17	48	.354
Pharo Village	76	141	.539
Warren	13	44	.295

TABLE 5.19

Numbers of Identified Specimens for the Most Abundant Taxon ($NISP_i$), Total Numbers of Identified Specimens ($\Sigma\,NISP$), and $NISP_i/\Sigma\,NISP$ for 13 Fremont Mammalian Species Assemblages

Site	$NISP_i$	$\Sigma\,NISP$	$NISP_i/\Sigma\,NISP$
Backhoe Village	39	100	.390
Bear River 1	1541	1859	.829
Bear River 2	786	943	.834
Bear River 3	670	827	.810
Caldwell Village	125	307	.407
Evans Mound	1963	4601	.427
Injun Creek	1713	1859	.921
Median Village	2341	2782	.841
Nephi Mound 1	636	781	.814
Old Woman	21	82	.256
Pharo Village	305	755	.404
Snake Rock	584	737	.792
Windy Ridge	20	37	.541

that within the Fremont mammalian species assemblages, diversity is correlated with sample size because $NISP_i/\Sigma\,NISP$ is correlated with sample size.

There are a number of possible explanations for this phenomenon. It seems most likely that the correlation is a small-sample effect. Eliminating the four smallest samples from the analysis (Caldwell Village, Backhoe Village, Old Woman, and Windy Ridge) also removes the significant correlation between $NISP_i/\Sigma\,NISP$ and $\Sigma\,NISP$ ($r_s = +.43$, $p > .20$) and between $1/\Sigma\,p_i^2$ and $\Sigma\,NISP$ ($r_s = -.42$, $p > .20$). However, correlation between diversity and sample size could also be a bone fragmentation effect. The more highly fragmented the bones of a given taxon in a given assemblage, the more specimens of that taxon there will be. Such fragmentation could be caused by butchering practices, and would in turn cause correlations between $NISP_i/\Sigma\,NISP$ and $\Sigma\,NISP$ across assemblages. It could also be that as Fremont peoples hunted more intensively, they increasingly focused that hunting on a single taxon. This, too, would cause correlations between $NISP_i/\Sigma\,NISP$ and $\Sigma\,NISP$, although in this case the diversity measure would actually be telling us about diversity. Because the cause of the correlation is unknown (although I suspect that the cause resides in the effects of small samples on relative abundance values), it would seem wise to be wary of assuming that the Fremont mammalian assemblage diversities are necessarily telling us about diversity. At least, however, they direct us toward the information needed to resolve the problem.

There is also a correlation between sample size and MNI-based diversity

TABLE 5.20
Minimum Numbers of Individuals for the Most Abundant Taxon (MNI_i), Total Minimum Numbers of Individuals (Σ MNI), and MNI_i/Σ MNI for Nine Fremont Avian Species Assemblages

Site	MNI_i	Σ MNI	MNI_i/Σ MNI
Bear River 1	10	63	.159
Bear River 3	10	74	.135
Evans Mound	2	12	.167
Injun Creek	11	44	.250
Knoll	7	.36	.194
Levee	18	139	.129
Nephi Mound 1	4	20	.200
Pharo Village	10	38	.263
Warren	3	24	.125

among the Fremont avian species assemblages (Table 5.17). Here, as sample size increases, Fremont avian assemblages become more diverse ($r_s = +.73$, $p < .05$). This is somewhat surprising. Since p_i^2 is calculated as MNI_i/Σ MNI, which in turn equals $aNISP^b_i/\Sigma\ aNISP^b$, then if $NISP_i/\Sigma$ NISP is correlated with sample size, the decrease in diversity could be a sample size effect. If so, however, it should not be a result of small sample sizes. All other things being equal, an MNI-based $1/\Sigma\ p_i^2$ should be higher, not lower, at small sample sizes. This is so because of the rapid decrease in MNI values as NISP values decrease (Chapter 2). At low sample sizes an MNI of, say, 50 spread across 20 taxa should produce low and roughly equivalent MNI counts for these taxa. Accordingly, low MNI counts should cause assemblages to appear even in terms of diversity, and thus to have high $1/\Sigma\ p_i^2$ values. But exactly the opposite has happened with the Fremont avian assemblages.

In fact, the correlation between sample size and MNI-based diversity values among the Fremont avian assemblages does not appear to be a sample size effect. Spearman's r_s between MNI_i/Σ MNI and Σ MNI, where i is again the most abundant species in the assemblage, is $-.267$ ($p > .20$; see Table 5.20), and I have already shown that there is no significant correlation between the $NISP_i/\Sigma$ NISP and Σ NISP values for the same assemblages.

Although the correlation is bothersome, the diversity values may well be measuring assemblage diversity. The five largest collections in terms of MNI also come from the five marsh-side sites (Bear River 1, Bear River 3, Levee, Knoll, and Injun Creek). Since marsh settings tend to be richer in bird species and individuals than non-marsh settings, it may well be that the sample size correlation I have found is an expectable result of the environmental setting of

5. TAXONOMIC RICHNESS, DIVERSITY, AND ASSEMBLAGE SIZE

the sites themselves. That is, the correlation might exist even if I were working with entire populations of bones from these sites. Unfortunately, since this is one of the facts that led Madsen to his hypothesis in the first place, accepting this explanation would prevent me from using these diversity values as a test of Madsen's hypothesis. Furthermore, I would feel much more secure about this explanation if I could feel secure that the Fremont avian samples are representative of the faunal populations from which they were drawn. Fremont archaeology, however, has generally been oriented toward the excavation of structures, and this strategy provides little basis for believing that our samples are in fact representative ones.

Although I believe I understand why correlations exist between sample size and diversity for the Fremont mammalian and MNI-based avian assemblages, I will not examine these values further. I am left with the Fremont NISP-based avian species diversity values, which are not significantly correlated with sample size.

According to Madsen's hypothesis, Fremont avian assemblage diversities should be higher in marsh-side than in non-marsh settings. In marsh settings, Madsen's argument goes, Fremont peoples utilized a wide range of marsh resources in generalist fashion. In other environmental settings, faunal and floral resources were secondary and these resources were not necessarily exploited in generalist fashion. Are the NISP-based avian species diversities in line with this hypothesis?

They are. Table 5.21 presents the avian species diversity values for the five marsh-side and three non-marsh sites in my sample (I have not used the Warren site because the published report on this site does not provide detailed information on its setting; see Enger and Blair 1947). As Madsen's hypothesis predicts, the average NISP-based avian species diversity of marsh-side sites is much higher than that of the non-marsh sites. Whether this supports the remainder of his argument is, of course, a very different matter, but at least the diversity indices support this aspect of his position.

I began my discussion of diversity indices by pointing out their potential value. Although my Fremont example is simplistic and was meant to provide an example of an approach rather than an actual analysis of Fremont adaptations themselves, I do think it illustrates the potential value of such indices in archaeological settings. It should, however, be clear that the application of diversity indices in archaeology is highly prone to sample size effects. Whether numbers of specimens or minimum numbers of individuals are used as the means of quantifying abundance, if $NISP_i / \Sigma\ NISP$ varies with $\Sigma\ NISP$, diversity values will also vary with $\Sigma\ NISP$. There are cases in which correlations between diversity measures and sample size may be a function of diversity relationships themselves, as I suspect occurs with the MNI-based $1/\Sigma\ p_i^2$ values for the Fremont avian species assemblages. However, unless the cause of the correlation can be

TABLE 5.21

Specimen-based Diversity Values ($1/\Sigma\ p_i^2$) for Fremont Avian Species Assemblages from Marsh and Non-Marsh Environmental Settings

Marsh-side sites	$1/\Sigma\ p_i^2$	Non-marsh-side sites	$1/\Sigma\ p_i^2$
Bear River 1	14.519	Evans Mound	5.548
Bear River 3	5.275	Nephi Mound 1	5.408
Injun Creek	3.352	Pharo Village	3.112
Knoll	5.475		
Levee	10.932		
$\bar{X}$	7.911		4.689

securely discovered, the meaning of such values becomes very unclear (see also Chapter 4). There are, of course, other, more standard taphonomic reasons why the meaning of diversity indices may be unclear. The fact that we are generally unable to infer modes of bone accumulation makes the application of diversity indices to assemblages from such sites as caves and rockshelters extremely hazardous, while the effects of bone fragmentation can leave us with as many questions as we began with. Nonetheless, even when the meaning of diversity indices is hazy, it is rare that they will not define the precise kinds of information that are needed to resolve their meaning, and this in itself is a major gain.

CHAPTER 6

Collection Techniques, Meat Weights, and Seasonality

In this chapter, I briefly discuss three issues that deal with very different aspects of archaeological faunal analysis: collection techniques, meat weights, and the analysis of the season(s) of the year during which an archaeological deposit accumulated. The first two of these topics have been very adequately discussed in the methodological literature, but the lessons of that literature have not been taken to heart by many practitioners. As a result, these matters bear repetition here. Although seasonality studies have also been the focus of much critical attention, that attention has not focused on nominal scale (presence/absence) analyses. Because there are hazards in such analyses, I will treat this topic as well.

Collection Techniques

Archaeologists are fully aware that their excavation and collection procedures play a major role in determining what they retrieve from the sites they excavate. Paleobotanists have paid much attention to the interrelationship between archaeological collection techniques and the nature of the plant macrofossils provided by those techniques (see the discussions and references in Ford 1979 and Hally 1981). Analysts dealing with faunal remains have also considered such interrelationships (e.g., Casteel 1972, 1976a, and references therein; Payne 1972b), but the widespread use of $\frac{1}{4}$-inch (.64-cm) screens in at least North American archaeology demonstrates that the results of these considerations have yet to be incorporated into standard archaeological collection procedures.

That screen-size choice can have a tremendous impact on the nature of the recovered fauna has been demonstrated by every study of this issue that has been done (see references above). Casteel (1972) used the excellent data

TABLE 6.1

Numbers of Identified Specimens Collected by Screen Size and by Body-Size Class From Three Nevada Sites[a]

Body-size class	Screen size (inches)		
	$\frac{1}{4}$	$\frac{1}{8}$	$\frac{1}{16}$
I	141	910	1930
II	626	1478	2450
III	1069	1358	1275
IV	85	4	0
V	1308	1	0

[a] From Thomas (1969:394).

provided by Thomas (1969) to investigate the magnitude of this impact, and I follow his lead here.

Thomas (1969) was interested in quantifying the loss of identifiable mammalian faunal materials that occurred as a result of the choice of various screen sizes. Accordingly, he collected vertebrates from three cave sites in northwestern Nevada by passing excavated sediments through a series of screens of $\frac{1}{4}$-inch (.64 cm), $\frac{1}{8}$-inch (.32 cm), and $\frac{1}{16}$-inch (.16 cm) size, and calculated the percentage of identifiable bone lost through $\frac{1}{4}$-inch and $\frac{1}{8}$-inch screens by assuming that no identifiable bones or teeth passed through his smallest screen. In order to examine the interrelationship between relative loss, screen size, and the size of the animals whose bones he was collecting, Thomas divided the mammalian species represented in the collection into five arbitrary size classes.

Class I: live weight less than 100 g (e.g., meadow mice),
Class II: live weight between 100 and 700 g (e.g., squirrels),
Class III: live weight between 700 g and 5 kg (e.g., cottontail rabbits),
Class IV: live weight between 5 and 25 kg (e.g., coyote and bobcat) and,
Class V: live weight greater than 25 kg (e.g., antelope, deer, and mountain sheep).

He was then able to present data on the number of bones of each mammal size class that was collected in each of his three screens.

Thomas (1969) presented his results in terms of six arbitrary levels, levels that preserved natural stratigraphic relationships within each sites. However, since I am interested only in bone loss, Thomas' analytic levels are not of concern, and I have merged all of Thomas' bone loss data into a single composite table. Table 6.1 presents that information, providing numbers of identified specimens recovered per body-size class per screen size for the combined data from all three

6. COLLECTION, MEAT WEIGHTS, AND SEASONALITY

TABLE 6.2
Percentage Recovery by Screen Size and Body-Size Class for Thomas' Three Nevada Faunas[a]

Body-size class	Screen size (inches)		
	$\frac{1}{4}$	$\frac{1}{8}$	$\frac{1}{16}$
I	05	31	65
II	14	33	54
III	29	37	34
IV	96	04	00
V	100	00	00

[a] From Thomas (1969:394).

sites. Table 6.2 transforms this information into percentage recovery data, showing the percentage of each body-size class recovered at each screen size.

There are a number of ways in which these recovery rates can be analyzed, but no matter how the analysis proceeds, the effects of screen size on recovery are seen to be dramatic.

Figure 6.1, for instance, plots body-size class against percentage recovered for $\frac{1}{4}$-inch screens. The rapid rise in recovery rate as body size increases is

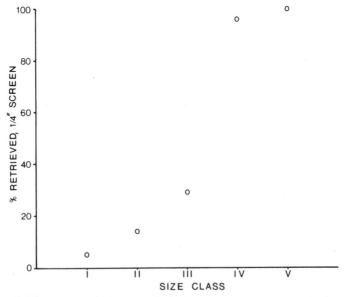

Figure 6.1 Percentage of identified specimens recovered per body-size class for Thomas' three Nevada faunas: $\frac{1}{4}$-inch (.64-cm) screen (From Thomas 1969.)

COLLECTION TECHNIQUES   171

obvious, and the biases that would result from analyzing the recovered fauna as a representative sample of what was there to recover should be clear. Figure 6.2 plots screen size against the cumulative percentage of bones and teeth retrieved for body-size Class I mammals. It is not surprising that of all the bones in the size class that were retrieved, only 5% were retrieved with a ¼-inch screen. Note, however, that only 31% of all identifiable Class I specimens were retrieved with a ⅛-inch screen. That is, even collecting with techniques that *far* surpass standard approaches in precision, nearly 75% of all identifiable small mammal remains were lost. Recall as well that this figure is based on the assumption that no identifiable specimens were lost through the 1/16-inch screen. Although this assumption was necessary for Thomas' analysis, the assumption was probably wrong, and his data thus must provide a conservative estimate of the loss of small mammal specimens through ¼-inch and ⅛-inch screen.

Thomas' important study reveals that even faunas from sites excavated as carefully as Gatecliff Shelter and Hidden Cave (for which only ⅛-inch screen was used) have been significantly modified as a result of collection techniques. One can only wonder about our understanding of the nature of Fremont faunas (see Chapter 5), since Fremont archaeology is routinely conducted with ¼-inch screens. The implication of Thomas' data seem obvious: both paleoenvironmental and subsistence (e.g., Stahl 1982) studies are probably routinely dealing with samples that are badly biased as a result of collection techniques *that are*

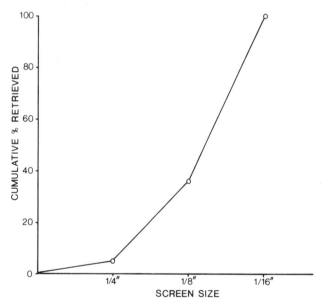

Figure 6.2 Cumulative percentage of identified specimens recovered by screen size for Thomas' three Nevada faunas: Class I mammals. (From Thomas 1969.)

under our own control. I cannot help but wonder why ¼-inch screens are still so popular among American archaeologists, although Payne (1972b) was certainly correct in pointing out that they are better than no screens at all.

Meat Weights

Archaeologists have frequently calculated the weight of meat represented per taxon in a faunal assemblage in order to assess the relative importance of those taxa in human subsistence. They have done this for a very simple reason. If bison and mice are each represented by a single specimen, and thus by a single individual, in an assemblage, both counts of specimens and counts of minimum numbers will treat those animals as equally frequent, although their contribution to the diet of the people involved would not have been so. Indeed, it was to assess subsistence contributions in terms of meat weight that White (1953) turned to the application of minimum numbers to archaeological faunas. In paleontology, meat weights have also been used to estimate the relative composition of faunas in terms of their biomass (e.g., Guthrie 1968).

Several methods are in use to calculate the weight of meat per taxon represented by the bones in a faunal aggregate. In each method, a measure thought to be correlated with meat weight is used to derive those weights through the application of some simple functional relationship. The commonly used measures are the weight of bone per taxon and the minimum number of individuals. In addition, measurements of single bones have been used to calculate the weights of the individuals from which those bones came.

When the weight of bone per taxon is used to derive meat weights, the analyst assumes that bone weight is a fixed percentage of meat weight. Bone weight is, therefore, multiplied by the appropriate factor to obtain meat weight (e.g., Cook and Treganza 1950; Kubasiewicz 1973; Reed 1963; Uerpmann 1973a,b). Traditionally, the major difficulty in applying this method has been thought to lie in the determination of the correct linear factor to use in converting bone weight to meat weight.

Casteel (1978), however, has shown that bone weights and meat weights in individuals are not related in a simple linear fashion, but are instead related curvilinearly. Thus, in order to estimate meat weight from bone weight, one may not use an equation of the form $Y = aX$, where Y is the predicted meat weight of the taxon involved; X, the weight of bone for that taxon; and a, a constant to be determined empirically. Instead, the actual relationship for individuals is of the form $Y = aX^b$, where both a and b are constants to be determined empirically.

Because the relationship between bone weight and meat weight for an individual is curvilinear, simple constants may not be used to convert one to the other. For instance, using the appropriate equation for pigs, Casteel (1978) showed that bone weight varied from 82% to 2% of meat weight (total tissue weight minus bone weight) as the weight of bone varied from 100 to 100,000,000 g. Casteel also noted that the curvilinear relationship between meat weight and bone weight applies to individuals, not to composite aggregates of faunal material. As a result, the equations may not be applied to such aggregates. Only when all the bones involved are from a single individual does it become appropriate to use an equation of this sort to predict meat weight from bone weight. One might assume that all the bones in the aggregate came from a single individual, but such an assumption is, of course, clearly inappropriate and generally demonstrably wrong. Casteel (1978) concluded, as did Chaplin (1971) before him, that this method of inferring meat weight from bone weight must be rejected.

If this method is inadequate, then the minimum number of individuals per taxon might be used to determine meat weights. In the most common application of this approach, the minimum number value for a given taxon is multiplied by the average weight of a modern individual of that taxon, and the resultant figures used in further analyses (e.g., Parmalee 1965; Parmalee and Klippel 1983; Smith 1975b; Stewart and Stahl 1977; White 1953; see Lyman 1979 for a thoughtful modification of this approach). It should be evident that meat weights so derived will suffer from the same problems that affect the minimum number of individual values on which they are based. In most cases, the approach can provide at best ordinal level data on the "dietary contribution" of taxa in terms of meat weights (Grayson 1979b; Lyman 1982b; see also Chapter 2).

If accurate meat weights for selected individuals are required, then a third approach might be employed. In this approach, equations are established that relate a measure of bone size (for instance, astragalus length) to meat weight for each taxon involved. Once these equations have been established, they may be used to predict the weights of individuals represented by archaeological or paleontological specimens. Noddle (1973) applied this approach to modern cattle, Casteel (1976b) has used it for various species of fish, and Reitz and Honerkamp (1983) have applied it to pig, sheep, and cattle. The advantages of this approach for estimating the weights of archaeological individuals are sizeable, since the resultant values are likely to be quite accurate, assuming the equations derived from modern vertebrates are applicable to their archaeological counterparts (Noddle 1973; see, however, the discussion in Reitz and Honerkamp 1983). Of course, this approach does not allow the analyst to determine meat weights for all individuals in a faunal assemblage, since there is usually no way of determining which specimens came from different animals. If the most abundant element is used to control for this problem, then difficulties relating to aggregation emerge. Although the method has the important potential of providing accurate meat weights for individual animals whose weight is of interest (e.g., Casteel 1974; Reitz and Honerkamp 1983), it does not provide a means of accurately assessing total meat weights per taxon in a faunal assemblage.

In short, there are two methods that have been commonly used to estimate the weight of meat per taxon represented by the bones within a faunal assemblage. Of these, the approach based on bone weight per taxon is invalid, while that based on minimum numbers of individuals can usually provide no better than ordinal scale information. A third approach to the estimation of meat weights, based on measurements of individual bone specimens, is potentially quite accurate, but can be applied only to individuals, not to entire assemblages. Thus, faunal analysts have no valid means of measuring taxonomic abundances within faunal assemblages in terms of meat weights or biomass above an ordinal scale.

Seasonality

Most attempts to infer the season or seasons during which a set of ancient deposits accumulated depend heavily on the kinds of organisms present in those deposits and on the state of maturity of those organisms (see the excellent review in Monks 1981). Such studies have an impressive antiquity; the British natural historian John Woodward (1665–1728) used precisely this approach in 1728 in an attempt to determine the season during which Noah's flood had occurred (Woodward 1728; see Grayson and Thomas 1983). Although there are many problems with efforts to infer seasonality (Monks 1981), here I wish to explore one difficulty that results from the fact that seasonality inferences drawn from taxonomic presences (and, at times, absences) assume that modern patterns of seasonal availability can be extended into the prehistoric past as an interpretive baseline. I will focus on the use of migratory birds, and in particular, migratory waterfowl, but I note that similar arguments can be applied to other vertebrates as well (Grayson and Thomas 1983).

It does not require deep thought to argue that modern abundances of species are often poor indicators of even the late prehistoric abundances of those species. The Passenger Pigeon *(Ectopistes migratorius)* provides an obvious example. It has been estimated that at the time of European contact, one of every three breeding birds in North America was a Passenger Pigeon, yet the last one died in 1914 (Schorger 1973). Clearly, the modern abundance of these birds tells us little about their abundances even a century ago.

Concern about the use of modern abundances in seasonality studies is relevant because seasonal frequency distributions of birds are generally unimodal or bimodal, with many individuals of a given species present during one or two seasons of the year and absent, or present in greatly reduced numbers, during

SEASONALITY

other seasons. In attempting to use presence/absence data for extracting seasonality information, archaeologists depend on the troughs in migratory bird seasonal frequency distributions, the months when the animals are now generally unavailable in the area of interest. If the troughs did not exist, the birds would be present year-round, and inferences concerning seasonality could not be drawn from their simple presence in a set of archaeological deposits. The difficulty, however, is that increasing the abundance of a given migratory bird as a whole increases the chances that a given seasonal trough will be occupied, while decreasing the abundance of that bird decreases that chance.

Take, for example, the frequencies of Common Goldeneyes *(Bucephala clangula)* across the year in the Lower Klamath National Wildlife Refuge of northeastern California. Figure 6.3 plots the number of Common Goldeneyes present in the refuge per week, across all 52 weeks of the year (data from the files of the Lower Klamath National Wildlife Refuge). The trough in this distribution occurs between the end of May and the beginning of November. An analyst with Common Goldeneyes in an archaeological assemblage from this area might use this bird to argue for an occupation between early November and late May. But note that Common Goldeneyes are never really very common in the Lower Klamath basin: the greatest number present at any one time during

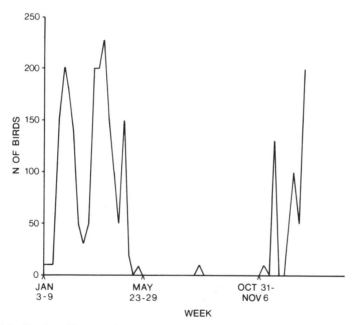

Figure 6.3 Number of Common Goldeneyes present by week during 1965 on the Lower Klamath National Wildlife Refuge, northeastern California.

1965 was less than 250. In fact, of the 23 species of waterfowl (ducks, geese, and swans) whose abundances have been monitored historically in the Lower Klamath Basin, only 5 are routinely absent. Of these 5, 4 are among the rarest waterfowl at any time of the year (in addition to Common Goldeneye, they are Trumpeter Swan [*Cygnus buccinator*], Ross' Goose [*Chen rossii*], and Wood Duck [*Aix sponsa*]). The rarer the bird, the less the chance that any individuals of that species will occupy the troughs in seasonal frequency distributions.

Conversely, the more abundant the bird, the greater the chance that a seasonal trough will be occupied. Figure 6.4 displays the weekly abundances of Redheads *(Aythya americana)* within the Lower Klamath Refuge during 1965. At no time during this year were Redheads completely absent from the Lower Klamath region, although only 20 were observed during one week in mid-January. Redheads, however, are often very abundant in this area: over 8000 individuals may be present during late summer. Were this bird reduced in numbers, the chances are good that the few individuals who now occupy the seasonal troughs would be eliminated. Although no archaeologist currently working in the Lower Klamath basin would be likely to use the presence of Redhead remains in an archaeological site as a seasonal indicator, the Redhead may well have been such an indicator in the past.

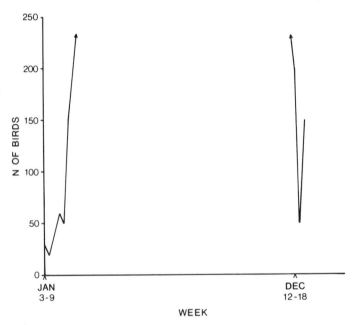

Figure 6.4 Number of Redheads present by week during 1965 on the Lower Klamath National Wildlife Refuge, northeastern California.

SEASONALITY

Because we have no way of knowing how the modern yearly frequency distribution of Common Goldeneyes compares to its prehistoric frequency distribution in the Lower Klamath basin, it would be hazardous to use this bird as a seasonal indicator. What we do know is that many migratory birds that are now uncommon were at one time much more abundant (e.g., Banko 1960). The more abundant they were, the less likely it is that they were absent during any season of the year.

Because we know so little about the past abundances of animals, we have little control over the applicability of modern patterns of seasonal presences and absences, and of seasonal activity as a whole, to the past. Clearly, it is the pattern of changing frequencies across seasons, and not seasonal absences, that contains the most secure information on seasonality. To examine such patterns, ordinal measures of taxonomic abundance, not simply presence/absence data, are needed. My point here, however, is that the inference of seasonality from the remains of migratory birds present in an archaeological fauna requires a major assumption about the abundances of those birds at the time the site was occupied. Everything we know about the history of bird populations during modern times suggests that this is an extremely hazardous assumption to make.

CHAPTER 7

Conclusions

I have had an opportunity of seeing a Cape hyaena at Oxford. . . . I was enabled also to observe the animal's mode of proceeding in the destruction of bones: the shin bone of an ox being presented to this hyaena, he began to bite off with his molar teeth large fragments from its upper extremity, and swallowed them whole as fast as they were broken off. On his reaching the medullary cavity, the bone split into angular fragments . . . he went on cracking it till he had extracted all the marrow . . . this done, he left untouched the lower condyle, which contains no marrow, and is very hard. The state and form of this residuary fragment are precisely like those of similar bones at Kirkdale; the marks of teeth on it are very few . . . these few, however, entirely resemble the impressions we find on the bones at Kirkdale; the small splinters also in form and size, and manner of fracture, are not distinguishable from the fossil ones . . . there is absolutely no difference between them, except in point of age.

William Buckland (1823:37–38)

The multitude of co-existing individuals is not to be reckoned from the absolute quantity of their fossil remains in a given locality. As reasonably might we infer the former populousness of a deserted village from the quantity of human bones in its churchyard.

Richard Owen (1846:xxvii)

Fossil remains indicate the comparative numbers of the animals to which they belonged, only when their destructibility and size are taken into consideration. Thus, the stone-like molars of the Mammoth would survive the destruction of all traces of the bones of the smaller animals, and remain in many instances as the sole evidence that Postglacial mammals ever dwelt in the area where they were found.

W. Boyd Dawkins (1869:207)

In order to account satisfactorily for the character presented by the ornithic fauna of the Caves of the Aquitanian region, it is necessary not only to draw up the list of the species which have been discovered there, but also to indicate the relative abundance of the bones of each; for one bird may be represented only by a single piece of the skeleton, while others will have left hundreds of remains. This study may supply valuable information: it shows us that certain species were evidently sought for by the inhabitants of the Caves; for their bones abound, and the delicacy of their flesh well justifies the pursuit to which they were exposed.

Alphonse Milne-Edwards (1875:226)

CONCLUSIONS

The issues that underlie the measurement problems that I have considered here, and the issues that taphonomists are now addressing, are not new. They developed as vertebrate paleontology and then prehistoric archaeology developed during the late eighteenth and nineteenth centuries (Grayson 1983a; Rudwick 1972). What is new is the number of people who are addressing these questions, and the detail in which they are being addressed.

But it is also true that taphonomic questions have been more vigorously pursued than have questions relating to the quantification of taxonomic abundances. The emphasis placed on taphonomy during the past decade or so is crucially important. As I have discussed, many issues relating to quantification initially arise as a result of taphonomic considerations. In addition, many quantitative investigations of taxonomic abundances provide us not with a set of convincing numbers, but instead with a set of precise questions that can only be addressed with equally precise and often unavailable taphonomic information.

Although the current emphasis on taphonomy is extremely healthy, the relative lack of concern with the meaning and behavior of our basic counting units is not. The problem is well illustrated by the fact that the most important work on cave taphonomy, and the most detailed analysis of the Kromdraai, Sterkfontein, and Swartkrans faunas, published to date is built on a one-paragraph discussion of quantification (Brain 1981). In North America, it is routine for even sizeable faunal analyses to be conducted without consideration of measurement issues, although here the problem is worse, since it is also common for North American faunal analyses to be conducted without concern for taphonomic questions.

Indeed, to some extent the problems I have discussed are general ones, modified only by the apparently peculiar nature of faunal data. In artifactual settings, the number of artifact classes in a series of tool assemblages can generally be tightly predicted from the sizes of those assemblages (Jones *et al.* 1983), just as the number of taxa within faunal assemblages can be tightly predicted from the number of identified bones and teeth contained by those assemblages. I would not be surprised to discover that the relative abundances of artifact classes across a series of tool assemblages are significantly correlated with the sizes of those assemblages, and thus would not be surprised to find that the artifact diversities of those assemblages are correlated with, and probably a function of, sample size. Schiffer's recent suggestion that the ratio of the "minimum number of vessels" (Schiffer 1983:687) to the total number of sherds in an archaeological assemblage be used as a measure in archaeological analysis provides another example. Although Schiffer (1983) seems to feel otherwise, it should be evident that such a measure will vary as a function of the number of sherds per assemblage.

My focus here, however, has been on the faunal record. If nothing else, I hope I have shown the importance of dealing with problems relating to how we count identified bones and teeth from archaeological sites, and of dealing with problems relating to the behavior of measurements derived from those counts. Even if the approaches I have suggested here prove to be flawed or simplistic or both, the problems themselves are real, and must be considered by all analysts working with vertebrate faunas from archaeological sites, just as those analysts must consider taphonomic processes.

References

AIKENS, C. M.
    1966   Fremont-Promontory-Plains relationships in northern Utah. *University of Utah Anthropological Papers* 82.
    1967   Excavations at Snake Rock Village and the Bear River No. 2 site. *University of Utah Anthropological Papers* 87.
    1970   Hogup Cave. *University of Utah Anthropological Papers* 93.

AMBLER, J. R.
    1966   Caldwell Village. *University of Utah Anthropological Papers* 84.

ANDREWS, P., and E. M. N. EVANS
    1983   Small mammal bone accumulations produced by mammalian carnivores. *Paleobiology* **9**:289–307.

BANKO, W. E.
    1960   The Trumpeter Swan: Its history, habits, and population in the United States. *North American Fauna* 63.

BATE, D. M. A.
    1937   Paleontology: The fossil fauna of the Wady El-Mughara Caves. In *The Stone Age of Mt. Carmel*, Volume 1, *Excavations at the Wady El-Mughara*, by D. A. E. Garrod and D. M. A. Bate. Oxford:Oxford University Press. Pp. 135–240.

BEDWELL, S. F.
    1969   Prehistory and environment of the pluvial Fort Rock Basin area of

south central Oregon. Ph.D. dissertation, Department of Anthropology, University of Oregon.
- 1973 *Fort Rock Basin: Prehistory and environment.* Eugene:University of Oregon Books.

BEHRENSMEYER, A. K.
- 1975 The taphonomy and paleoecology of Plio-Pleistocene vertebrate assemblages east of Lake Rudolf, Kenya. *Museum of Comparative Zoology, Harvard University, Bulletin* 146:473–578.
- 1978 Taphonomic and ecologic information from bone weathering. *Paleobiology* **4**:150–162.

BEHRENSMEYER, A. K., and D. E. B. BOAZ
- 1980 The recent bones of Amboseli National Park, Kenya, in relation to East African paleoecology. In *Fossils in the making*, edited by A. K. Behrensmeyer and A. P. Hill. Chicago:University of Chicago Press. Pp. 72–92.

BEHRENSMEYER, A. K., and A. P. HILL (Editors)
- 1980 *Fossils in the making.* Chicago:University of Chicago Press.

BERRY, M. S.
- 1972 *The Evans Site.* Special Report, Department of Anthropology, University of Utah.

BINFORD, L. R.
- 1978 *Nunamiut ethnoarchaeology.* New York:Academic Press.
- 1981 *Bones: Ancient men and modern myths.* New York:Academic Press.
- 1984 *Faunal remains from Klasies River mouth.* New York:Academic Press.

BINFORD, L. R., and J. B. BERTRAM
- 1977 Bone frequencies—and attritional processes. In *For theory building in archaeology*, edited by L. R. Binford. New York:Academic Press. Pp. 77–153.

BÖKÖNYI, S.
- 1970 A new method for the determination of the minimum number of individuals in animal bone material. *American Journal of Archaeology* **74**:291–292.

BRADLEY, J. V.
- 1968 *Distribution-free statistical tests.* Englewood-Cliffs:Prentice-Hall.

BRAIN, C. K.
- 1967 Hottentot food remains and their bearing on the interpretation of fossil bone assemblages. *Scientific Papers of the Namib Desert Research Station* **32**:1–11.
- 1969 The contribution of Namib Desert Hottentots to an understanding

of australopithecine bone accumulations. *Scientific Papers of the Namib Desert Research Station* **39**:13-22.

1976 Some principles in the interpretation of bone accumulations associated with man. In *Human origins and the East African evidence*, edited by G. L. Isaac and E. R. McCown. Menlo Park:Staples Press. Pp. 96-116.

1981 *The hunters or the hunted? An introduction to African cave taphonomy*. Chicago:University of Chicago Press.

BROWN, J. H.

1971 Mammals on mountaintops: Nonequilibrium insular biogeography. *American Naturalist* **105**:467-478.

1978 The theory of insular biogeography and the distribution of boreal birds and mammals. In Intermountain biogeography, edited by K. T. Harper and J. L. Reveal. *Great Basin Naturalist Memoirs* **2**:209-228.

BUCKLAND, W.

1823 *Reliquiae diluvianae, or, observations on the organic remains contained in caves, fissures, and diluvial gravel, and on other geological phenomena, attesting to the action of an universal deluge*. London:Murray.

BUTLER, B. R.

1972 The Holocene or postglacial ecological crisis on the eastern Snake River plain. *Tebiwa* **15**:(1):49-61.

CARLISLE, R. C., and J. M. ADOVASIO (editors)

1982 *Meadowcroft: Collected papers on the archaeology of Meadowcroft Rockshelter and the Cross Creek Drainage*. Department of Anthropology, University of Pittsburgh.

CASTEEL, R. W.

1972 Some biases in the recovery of archaeological faunal remains. *Proceedings of the Prehistoric Society* **36**:382-388.

1974 Use of Pacific salmon otoliths for estimating fish size, with a note on the size of late Pleistocene and Pliocene salmonids. *Northwest Science* **48**:175-179.

1976a Comparison of column and whole unit samples for recovering fish remains. *World Archaeology* **8**:192-196.

1976b *Fish remains in archaeology and paleo-environmental studies*. New York:Academic Press.

1976/1977 A consideration of the behaviour of the minimum number of individuals index: A problem in faunal characterization. *Ossa* **3/4**:141-151.

1977 Characterization of faunal assemblages and the minimum number

of individuals determined from paired elements: Continuing problems in archaeology. *Journal of Archaeological Science* **4**:125–134.
1978   Faunal assemblages and the "weigemethode" or weight method. *Journal of Field Archaeology* **5**:71–77.

CHAPLIN, R. E.
1971   *The study of animal bones from archaeological sites.* New York:Seminar Press.

CLELAND, C. E.
1966   The prehistoric animal ecology and ethnozoology of the upper Great Lakes region. *Museum of Anthropology, University of Michigan, Anthropological Papers* 29.

CODY, M. L.
1974   Competition and the structure of bird communities. *Monographs in Population Biology* 7.

CODY, M. L., and J. M. DIAMOND (Editors)
1975   *Ecology and evolution of communities.* Cambridge:Belknap Press.

CONOVER, W. J.
1971   *Practical nonparametric statistics.* New York:Wiley.

COOK, S. F., and A. E. TREGANZA
1950   The quantitative investigation of Indian mounds. *University of California Publications in American Archaeology and Ethnology* **40**:223–262.

COWGILL, G. L.
1977   The trouble with significance tests and what we can do about it. *American Antiquity* **42**:350–368.

DALY, P.
1969   Approaches to faunal analysis in archaeology. *American Antiquity* **34**:146–153.

DAMUTH, J.
1982   Analysis of the preservation of community structure in assemblages of fossil mammals. *Paleobiology* **8**:434–446.

DAVIS, J. O.
1984   Sediments and geological setting of Hidden Cave. In The archaeology of Hidden Cave, edited by D. H. Thomas. *American Museum of Natural History Anthropological Papers*, in press.

DAVIS, J. O., W. N. MELHORN, D. F. TREXLER, and D. H. THOMAS
1983   Geology of Gatecliff Shelter: Physical stratigraphy. In The archaeology of Monitor Valley: 2. Gatecliff Shelter, by D. H. Thomas. *American Museum of Natural History Anthropological Papers* **59**(1):39–63.

DAVIS, M. B.
　1967　On the theory of pollen analysis. *American Journal of Science* **261**:897-912.
DAWKINS, W. B.
　1869　On the distribution of British postglacial mammals. *Quarterly Journal of the Geological Society of London* **25**:192-217.
DODSON, P.
　1971　Sedimentology and taphonomy of the Oldman Formation (Campanian), Dinosaur Provincial Park, Alberta (Canada). *Palaeogeography, Palaeoclimatology, Palaeoecology* **10**:21-74.
DRAPER, N. R., and H. SMITH
　1966　*Applied regression analysis.* New York:Wiley.
DUCOS, P.
　1968　L'origine des animaux domestiques en Palestine. *Publications de l'Institut de Préhistoire de l'Université de Bordeaux*, Mémoire 6.
DUNNELL, R. C.
　1972　The prehistory of Fishtrap, Kentucky. *Yale University Publications in Anthropology* 75.
　1982　Americanist archaeological literature: 1981. *American Journal of Archaeology* **86**:509-529.
ENGER, W. D., Jr., and W. BLAIR
　1947　Crania from the Warren Mounds and their possible significance to Northern Periphery archaeology. *American Antiquity* **13**:142-146.
ESTES, R., and P. BERBERIAN
　1970　Paleoecology of a late Cretaceous vertebrate community from Montana. *Breviora* **343**:1-35.
EVERITT, B. S.
　1977　*The analysis of contingency tables.* London:Chapman and Hall.
FIELLER, N. R. J., and A. TURNER
　1982　Number estimation in vertebrate samples. *Journal of Archaeological Science* **9**:49-62.
FLANNERY, K. V.
　1967　The vertebrate fauna and hunting patterns. In *The Prehistory of the Tehuacan Valley*, Volume 1, edited by D. S. Byers. Austin:University of Texas Press. Pp. 132-178.
FORD, R. I.
　1979　Paleoethnobotany in American archaeology. *Advances in archaeological method and theory* **2**:285-336.
GEJVALL, N.-G.
　1969　*Lerna, a pre-Classical site in the Argoilid*, Volume 1, *The fauna.* Princeton:American School of Classical Studies at Athens.

GIFFORD, D. P.
- 1981 Taphonomy and paleoecology: A critical review of archaeology's sister disciplines. *Advances in archaeological method and theory* **4**:365–438.

GIFFORD, D. P., G. L. ISAAC, and C. M. NELSON
- 1980 Evidence for predation and pastoralism at Prolonged Drift: A Pastoral Neolithic site in Kenya. *Azania* **15**:57–108.

GILBERT, A. S., and B. H. SINGER
- 1982 Reassessing zooarchaeological quantification. *World Archaeology* **14**:21–40.

GILBERT, A. S., and P. STEINFELD
- 1977 Faunal remains from Dinkha Tepe, northwestern Iran. *Journal of Field Archaeology* **4**:329–351.

GOSS, R. S.
- 1983 *Deer antlers: Regeneration, function, and evolution.* New York:Academic Press.

GRAYSON, D. K.
- 1973 On the methodology of faunal analysis. *American Antiquity* **38**:432–439.
- 1974a Minimum numbers and sample size in vertebrate faunal analysis. Paper presented at the 39th Annual Meeting of the Society for American Archaeology, Washington, D.C.
- 1974b The Riverhaven No. 2 vertebrate fauna: Comments on methods in faunal analysis and on aspects of the subsistence potential of prehistoric New York. *Man in the Northeast* **8**:23–39.
- 1976 The Nighfire Island avifauna and the Altithermal. In Holocene environmental change in the Great Basin, edited by R. Elston. *Nevada Archaeological Survey Research Reports* **6**:74–102.
- 1977a On the Holocene history of some Northern Great Basin lagomorphs. *Journal of Mammalogy* **58**:507–513.
- 1977b Paleoclimatic implications of the Dirty Shame Rockshelter mammalian fauna. *Tebiwa: Miscellaneous Papers of the Idaho State University Museum of Natural History* 9.
- 1978a Minimum numbers and sample size in vertebrate faunal analysis. *American Antiquity* **43**:53–65.
- 1978b Reconstructing mammalian communities: A discussion of Shotwell's method of paleoecological analysis. *Paleobiology* **4**:77–81.
- 1979a Mt. Mazama, climatic change, and Fort Rock basin archaeofaunas. In *Volcanic activity and human ecology*, edited by P. D. Sheets and D. K. Grayson, New York:Academic Press. Pp. 427–458.
- 1979b On the quantification of vertebrate archaeofaunas. *Advances in archaeological method and theory* **2**:199–237.

1981a A mid-Holocene record for the heather vole, *Phenacomys* cf. *intermedius*, in the central Great Basin and its paleoecological significance. *Journal of Mammalogy* **62**:115–121.

1981b The effects of sample size on some derived measures in vertebrate faunal analysis. *Journal of Archaeological Science* **8**:77–88.

1982 Toward a history of Great Basin mammals during the past 15,000 years. In Man and environment in the Great Basin, edited by D. B. Madsen and J. F. O'Connell. *Society for American Archaeology Papers* **2**:82–101.

1983a *The establishment of human antiquity.* New York:Academic Press.

1983b The paleontology of Gatecliff Shelter: Small mammals. In The archaeology of Monitor Valley: 2. Gatecliff Shelter, by D. H. Thomas. *American Museum of Natural History Anthropological Papers* **59**(1):99–129.

1984 The paleontology of Hidden Cave: Birds and mammals. In The archaeology of Hidden Cave, edited by D. H. Thomas. *American Museum of Natural History Anthropological Papers*, in press.

GRAYSON, D. K., and D. H. THOMAS

1983 Seasonality at Gatecliff Shelter. In The archaeology of Monitor Valley: 2. Gatecliff Shelter, by D. H. Thomas. *American Museum of Natural History Anthropological Papers* **59**(1):434–438.

GREENE, J. L., and T. W. MATTHEWS

1976 Faunal study of unworked mammalian bones. In *The Hohokam: Desert farmers and craftsmen,* by E. W. Haury. Tucson:University of Arizona Press. Pp. 367–373.

GUILDAY, J. E.

1970 Animal remains from archaeological excavations at Fort Ligonier. *Annals of the Carnegie Museum* **42**:177–186.

1971 Biological and archaeological analysis of bones from a 17th century Indian Village (46 Pu 31), Putnam County, West Virginia. *West Virginia Geological and Economic Survey Report of Archaeological Investigations* 4.

GUILDAY, J. E., H. W. HAMILTON, E. ANDERSON, and P. W. PARMALEE

1978 The Baker Bluff Cave deposit, Tennessee, and the late Pleistocene faunal gradient. *Carnegie Museum of Natural History Bulletin* 11.

GUILDAY, J. E., and P. W. PARMALEE

1982 Vertebrate faunal remains from Meadowcroft Rockshelter, Washington County, Pennsylvania: A re-evaluation and interpretation. In *Meadowcroft: Collected papers on the archaeology of Meadowcroft Rockshelter and the Cross Creek drainage,* edited by R. C.

Carlisle and J. M. Adovasio. Department of Anthropology, University of Pittsburgh. Pp. 163–174.

GUILDAY, J. E., P. W. PARMALEE, and H. W. HAMILTON
  1977 The Clark's Cave bone deposit and the late Pleistocene paleoecology of the central Appalachian mountains of Virginia. *Carnegie Museum of Natural History Bulletin* 2.

GUILDAY, J. E., P. W. PARMALEE, and D. P. TANNER
  1962 Aboriginal butchering techniques at the Eschelman site (36 La 12), Lancaster County, Pennsylvania. *Pennsylvania Archaeologist* **32**:59–83.

GUILDAY, J. E., P. W. PARMALEE, and R. C. WILSON
  n.d. *Vertebrate faunal remains from Meadowcroft Rockshelter (36WH297), Washington County, Pennsylvania.* Manuscript on file, Department of Anthropology, University of Pittsburgh, Pittsburgh.

GUTHRIE, R. D.
  1967 Differential preservation and recovery of Pleistocene large mammal remains in Alaska. *Journal of Paleontology* **41**:243–246.
  1968 Paleoecology of the large-mammal community in interior Alaska during the late Pleistocene. *American Midland Naturalist* **79**:346–363.

HALL, E. R.
  1946 *Mammals of Nevada.* Berkeley:University of California Press.

HALLY, D. J.
  1981 Plant preservation and the content of paleobotanical samples: A case study. *American Antiquity* **46**:723–742.

HARPER, K. T., and G. M. ALDER
  1970 The macroscopic plant remains of the deposits of Hogup Cave and their paleoclimatic interpretation. In Hogup Cave, by C. M. Aikens. *University of Utah Anthropological Papers* **93**:215–240.

HARRIS, A. H.
  1977 Wisconsin age environments in the northern Chihuahuan Desert: Evidence from the higher vertebrates. In Transactions: Symposium on the biological resources of the Chihuahuan Desert, U.S. and Mexico, edited by R. H. Wauer and D. H. Riskind. *National Park Service Transactions and Proceedings Series* **13**:23–51.

HAURY, E.W.
  1976 *The Hohokam: Desert farmers and craftsmen.* Tucson:University of Arizona Press.

HESSE, B.
  1982 Bias in the zooarchaeological record: Suggestions for interpretation of bone counts in faunal samples from the Plains. In Plains

REFERENCES

Indian Studies: A collection of essays in honor of John C. Ewers and Waldo R. Wedel, edited by D. H. Ubelaker and H. J. Viola. *Smithsonian Contributions to Anthropology* **30**:157-172.

HESSE, B., and D. PERKINS, Jr.
1974 Faunal remains from Karataş-Semayük in southeast Anatolia: An interim report. *Journal of Field Archaeology* **1**:149-160.

HOLMER, R. N., and D. G. WEDER
1980 Common post-Archaic projectile points of the Fremont area. In Fremont perspectives, edited by D. B. Madsen. *Antiquities Section Selected Papers* (Utah Division of State History) **7**(16):55-68.

HOLTZMAN, R. C.
1979 Maximum likelihood estimation of fossil assemblage composition. *Paleobiology* **5**:77-89.

HOWARD, H.
1930 A census of the Pleistocene birds of Rancho La Brea from the collections of the Los Angeles Museum. *Condor* **32**:81-88.

HURLBERT, S. H.
1971 The nonconcept of species diversity: A critique and alternative parameters. *Ecology* **52**:577-586.

JELINEK, A. J.
1982 The Tabun Cave and Paleolithic man in the Levant. *Science* **216**:1369-1375.

JENNINGS, J. D
1974 *Prehistory of North America.* Second edition. New York:McGraw-Hill.
1978 Prehistory of Utah and the eastern Great Basin. *University of Utah Anthropological Papers* 98.

JOLICOEUR, P.
1963 Bilateral symmetry and asymmetry in limb bones of *Martes americana* and man. *Revue Canadienne de Biologie* **22**:409-432.

JONES, G. T., D. K. GRAYSON, and C. BECK
1983 Artifact class richness and sample size in archaeological surface assemblages. In Lulu Linear Punctated: Essays in honor of George Irving Quimby, edited by R. C. Dunnell and D. K. Grayson. *Museum of Anthropology, University of Michigan, Anthropological Papers* **72**:55-73.

KELLY, R. L., and E. HATTORI
1984 Environmental and historical background. In The archaeology of Hidden Cave, edited by D. H. Thomas. *American Museum of Natural History Anthropological Papers*, in press.

KENDALL, M. G.
1970 *Rank correlation methods.* London:Griffin.

KLEIN, R. G.
- 1975 Middle Stone Age man–animal relationships in southern Africa: Evidence from Die Kelders and Klasies River Mouth. *Science* **190**:265–277.
- 1976 The mammalian fauna of the Klasies River Mouth sites, southern Cape Province, South Africa. *South African Archaeological Bulletin* **31**:75–98.
- 1977 The ecology of early man in South Africa. *Science* **197**:115–126.
- 1978 Stone age predation on large African bovids. *Journal of Archaeological Science* **5**:195–217.
- 1980 Environmental and ecological implications of large mammals from Upper Pleistocene and Holocene sites in southern Africa. *Annals of the South African Museum* **81**(7).
- 1981 Stone age predation on small African bovids. *South African Archaeological Bulletin* **36**:55–65.
- 1982 Age (mortality) profiles as a means of distinguishing hunted species from scavenged ones in Stone Age archaeological sites. *Paleobiology* **8**:151–158.
- 1983 The Stone Age prehistory of southern Africa. *Annual Review of Anthropology* **12**:25–48.

KRANTZ, G. S.
- 1968 A new method of counting mammal bones. *American Journal of Archaeology* **72**:286–288.

KUBASIEWICZ, M.
- 1973 Spezifische Elemente der polnischen archäozoologischen Forschungen des letzen Vierteljahrhunderts. In *Domestikationsforschung und Geschichte der Haustiere*, edited by J. Matolsci. Budapest:Akadémiai Kiadó. Pp. 371–376.

LYMAN, R. L.
- 1979 Available meat from faunal remains: A consideration of techniques. *American Antiquity* **44**:536–546.
- 1982a Archaeofaunas and subsistence studies. *Advances in archaeological method and theory* **5**:331–393.
- 1982b The taphonomy of vertebrate archaeofaunas: Bone density and differential survivorship of fossil classes. Ph.D. dissertation, Department of Anthropology, University of Washington.

LYMAN, R. L., and S. D. LIVINGSTON
- 1983 Late Quaternary mammalian zoogeography of eastern Washington. *Quaternary Research* **20**:360–373.

MACARTHUR, R. H.
- 1972 *Geographical ecology: Patterns in the distribution of species.* New York:Harper and Row.

REFERENCES

MADSEN, D. B.
- 1975 Three Fremont sites in Emery County, Utah. *Antiquities Section Selected Papers* (Utah Division of State History) **1**(1).
- 1980a Fremont perspectives. *Antiquities Section Selected Papers* (Utah Division of State History) **7**(16).
- 1980b Fremont/Sevier subsistence. In Fremont perspectives, edited by D. B. Madsen. *Antiquities Section Selected Papers* (Utah Division of State History) **7**(16):25-34.
- 1982 Get it where the gettin's good: A variable model of Great Basin subsistence and settlement based on data from the eastern Great Basin. In Man and environment in the Great Basin, edited by D. B. Madsen and J. F. O'Connell. *Society for American Archaeology Papers* **2**:202-226.

MADSEN, D. B., and L. W. LINDSAY
- 1977 Backhoe Village. *Antiquities Section Selected Papers* (Utah Division of State History) **4**(12).

MARWITT, J. P.
- 1968 Pharo Village. *University of Utah Anthropological Papers* 91.
- 1970 Median Village and Fremont culture regional variation. *University of Utah Anthropological Papers* 95.

MAY, R. M.
- 1975 Patterns of species abundance and diversity. In *Ecology and evolution of communities*, edited by M. L. Cody and J. M. Diamond. Cambridge:Belknap Press. Pp. 81-120.

MEAD, J. I., R. S. THOMPSON, and T. R. VAN DEVENDER
- 1982 Late Wisconsinan and Holocene fauna from Smith Creek Canyon, Snake Range, Nevada. *San Diego Society of Natural History Transactions* **20**(1):1-26.

MEHRINGER, P. J., Jr.
- 1977 Great Basin late Quaternary environments and chronology. In Models and Great Basin prehistory, edited by D. D. Fowler. *Desert Research Institute Publications in the Social Sciences* **12**:113-167.

MERRIAM, J. C., and C. STOCK
- 1932 The Felidae of Rancho La Brea. *Carnegie Institute of Washington Publication* 422.

MILNE-EDWARDS, H.
- 1875 Observations on the birds whose bones have been found in the caves of the south-west of France. In *Reliquiae Aquitanicae; being contributions to the archaeology and paleontology of Perigord and the adjoining provinces of southern France*, by E. Lartet and H. Christy and edited by T. R. Jones. London:Williams and Norgate. Pp. 226-247.

MONKS, G. G.
    1981   Seasonality studies. *Advances in archaeological method and theory* **4**:177–240.

MORRISON, R. B.
    1964   Lake Lahontan: Geology of southern Carson Desert, Nevada. *United States Geological Survey Professional Paper* 401.

NICHOL, R. K., and G. A. CREAK
    1979   Matching paired elements among archaeological bone remains. *Newsletter of Computer Archaeology* **14**:6–16.

NODDLE, B. A.
    1973   Determination of the body weight of cattle from bone measurements. In *Domestikationsforschung und Geschichte der Haustiere*, edited by J. Matolsci. Budapest:Akadémiai Kiadó. Pp. 377–389.

OWEN, R.
    1846   *A history of British fossil mammals, and birds*. London:Van Voorst.

PARMALEE, P. W.
    1959   Animal remains from Raddatz Rockshelter, SK5, Wisconsin. *Wisconsin Archaeologist* **4**:83–90.
    1965   The food economy of Archaic and Woodland peoples at the Tick Creek Cave site, Missouri. *Missouri Archaeologist* **27**:1–34.
    1977   The avifauna from prehistoric sites in South Dakota. *Plains Anthropologist* **22**:189–222.
    1980   Utilization of birds by the Archaic and Fremont cultural groups of Utah. In Papers in avian paleontology honoring Hildegarde Howard, edited by K. E. Campbell, Jr. *Natural History Museum of Los Angeles County Contributions in Science* **330**:237–250.

PARMALEE, P. W., and W. E. KLIPPEL
    1983   The role of native animals in the food economy of the historic Kickapoo in central Illinois. In Lulu Linear Punctated: Essays in honor of George Irving Quimby, edited by R. C. Dunnell and D. K. Grayson. *Museum of Anthropology, University of Michigan, Anthropological Papers* **72**:253–324.

PARMALEE, P. W., A. A. PALOUMPIS, and N. WILSON
    1972   Animals utilized by Woodland peoples occupying the Apple Creek site, Illinois. *Illinois State Museum Reports of Investigations* 23.

PAYNE, S.
    1972a   On the interpretation of bone samples from archaeological sites. In *Papers in economic prehistory*, edited by E. S. Higgs. London:Cambridge University Press. Pp. 65–81.
    1972b   Partial recovery and sample bias: The results of some sieving ex-

periments. In *Papers in economic prehistory*, edited by E. S. Higgs. London:Cambridge University Press. Pp. 49–64.

PEET, R. K.
    1974   The measurement of species diversity. *Annual Review of Ecology and Systematics* **5**:285–307.

PERKINS, D., Jr., and P. DALY
    1968   A hunter's village in Neolithic Turkey. *Scientific American* **219**(5):96–106.

PIELOU, E. C.
    1975   *Ecological diversity*. New York:Wiley.

RAUP, D. M.
    1977   Species diversity in the Phanerozoic: Systematists follow the fossils. *Paleobiology* **3**:328–329.

READ-MARTIN, C. E., and D. W. READ
    1975   Australopithecine scavenging and human evolution: An approach from faunal analysis. *Current Anthropology* **16**:359–368.

REED, C. A.
    1963   Osteo-archaeology, In *Science in archaeology*, edited by D. Brothwell and E. S. Higgs. London:Thames and Hudson. Pp. 204–216.

REITZ, E., and N. HONERKAMP
    1983   British colonial subsistence strategy on the southeastern coastal plain. *Historical Archaeology* **17**:4–26.

RUDWICK, M. J. S.
    1972   *The meaning of fossils*. New York:American Elsevier.

SANDERS, H. L.
    1968   Marine benthic diversity: A comparative study. *American Naturalist* **102**:243–282.

SCHIFFER, M. B.
    1983   Toward the identification of formation processes. *American Antiquity* **48**:675–706.

SCHORGER, A. W.
    1973   *The Passenger Pigeon: Its natural history and exinction*. Norman:University of Oklahoma Press.

SCHROEDL, A. R., and P. F. HOGAN
    1975   Innocents Ridge and the San Rafael Fremont. *Antiquities Section Selected Papers* (Utah Division of State History) **1**(2).

SHARROCK, F. W., and J. P. MARWITT
    1967   Excavations at Nephi, Utah, 1965–1966. *University of Utah Anthropological Papers* 88.

SHEEHAN, P. M.
    1977   Species diversity in the Phanerozoic: A reflection of labor by systematists? *Paleobiology* **3**:325–328.

SHIELDS, W. F.
- 1967 1966 excavations: Uinta Basin. *University of Utah Anthropological Papers* **89**:1–32.

SHIELDS, W. F., and G. F. DALLEY
- 1978 The Bear River No. 3 site. *University of Utah Anthropological Papers* **99**:55–104.

SHIPMAN, P.
- 1981 *Life history of a fossil: An introduction to taphonomy and paleoecology.* Cambridge:Harvard University Press.

SHOTWELL, J. A.
- 1955 An approach to the paleoecology of mammals. *Ecology* **36**:327–337.
- 1958 Inter-community relationships in Hemphillian (mid-Pliocene) mammals. *Ecology* **39**:271–282.
- 1963 The Juntura Basin: Studies in earth history and paleoecology. *Transactions of the American Philosophical Society* **53**(1).

SIMPSON, G. G.
- 1965 *The geography of evolution.* New York:Capricorn Books.

SMITH, B. D.
- 1975a Middle Mississippi exploitation of animal populations. *Museum of Anthropology, University of Michigan, Anthropological Papers* 57.
- 1975b Toward a more accurate estimation of the meat yield of animal species at archaeological sites. In *Archaeozoological studies*, edited by A. T. Clason. Amsterdam:North-Holland Publishing. Pp. 99–106.
- 1979 Measuring the selective utilization of animal species by prehistoric human populations. *American Antiquity* **44**:155–160.

SPAULDING. W. G., E. B. LEOPOLD, and T. R. VAN DEVENDER
- 1983 Late Wisconsin paleoecology of the American Southwest. In *The late Pleistocene environments of the United States*, edited by S. C. Porter. Minneapolis:University of Minnesota Press. Pp. 259–293.

STAHL, P. W.
- 1982 On small mammal remains in archaeological context. *American Antiquity* **47**:822–829.

STEWART F. L., and P. W. STAHL
- 1977 Cautionary note on edible meat poundage figures. *American Antiquity* **42**:267–270.

STOCK, C.
- 1929 A census of the Pleistocene mammals of Rancho La Brea, based on the collections of the Los Angeles Museum. *Journal of Mammalogy* **10**:281–289.

REFERENCES

STYLES, B. W.
    1981   Faunal exploitation and resource selection: Early Late Woodland subsistence in the lower Illinois valley. *Northwestern University Archaeological Program Scientific Papers* 3.

TAYLOR, D. C.
    1957   Two Fremont sites and their position in Southwestern prehistory. *University of Utah Anthropological Papers* 29.

THOMAS, D. H.
    1969   Great Basin hunting patterns: A quantitative method for treating faunal remains. *American Antiquity* **34**:393–401.
    1971   On distinguishing natural from cultural bone in archaeological sites. *American Antiquity* **36**:366–371.
    1983a The archaeology of Monitor Valley: 1. Epistemology. *American Museum of Natural History Anthropological Papers* **58**(1).
    1983b The archaeology of Monitor Valley: 2. Gatecliff Shelter. *American Museum of Natural History Anthropological Papers* **59**(1).
    1984   Previous research at Hidden Cave. In The archaeology of Hidden Cave, edited by D. H. Thomas. *American Museum of Natural History Anthropological Papers*, in press.

THOMAS, D. H., and D. MAYER
    1983   Behavioral faunal analysis of selected horizons. In The archaeology of Monitor Valley: 2. Gatecliff Shelter, by David H. Thomas. *American Museum of Natural History Anthropological Papers* **59**(1):353–391.

THOMAS, D. H., and D. PETER
    1984   Excavation strategies at Hidden Cave: 1940–1980. In The archaeology of Hidden Cave, edited by D. H. Thomas. *American Museum of Natural History Anthropological Papers*, in press.

THOMPSON, R. S.
    1983   Modern vegetation and climate. In The archaeology of Monitor Valley: 1. Epistemology, by D. H. Thomas. *American Museum of Natural History Anthropological Papers* **58**(1):99–106.

THOMPSON, R. S., and J. I. MEAD
    1982   Late Quaternary environments and biogeography of the Great Basin. *Quaternary Research* **17**:39–55.

TIPPER, J. C.
    1979   Rarefaction and rarefiction—the use and abuse of a method in paleoecology. *Paleobiology* **5**:423–434.

UERPMANN, H.-P.
    1973a Animal bone finds and economic archaeology: A critical study of "osteoarchaeological" method. *World Archaeology* **4**:307–322.
    1973b Ein Beitrag zur Methodik der wirtschaftshistorischen Auswertung

von Tierknochenfunden aus Siedlungen. In *Domestikationsforschung und Geschichte der Haustiere,* edited by J. Matolsci. Budapest:Akadémiai Kiadó. Pp. 391–396.

VAN DEVENDER, T. R., and W. G. SPAULDING
 1979 Development of vegetation and climate in the southwestern United States. *Science* **204**:701–710.

VAUFREY, R.
 1939 Paléolithique et Mésolithique palestiniens. *Revue Scientifique* **77**:390–406.

VOORHIES, M. R.
 1969 Taphonomy and population dynamics of an early Pliocene vertebrate fauna, Knox County, Nebraska. *University of Wyoming Contributions to Geology Special Papers* **1**:1–69.

WATSON, J. P. N.
 1972 Fragmentation analysis of animal bone samples from archaeological sites. *Archaeometry* **14**:221–227.
 1979 The estimation of the relative frequencies of mammalian species: Khirokitia 1972. *Journal of Archaeological Science* **6**:127–137.

WHEAT, J. B.
 1972 The Olsen–Chubbock site: A Paleo-Indian bison kill. *Society for American Archaeology Memoirs* **26**.

WHITE, T. E.
 1953 A method of calculating the dietary percentages of various food animals utilized by aboriginal peoples. *American Antiquity* **18**:396–398.

WIGAND, P. E., and P. J. MEHRINGER, Jr.
 1984 Plant and seed analyses for Hidden Cave. In The archaeology of Hidden Cave, edited by D. H. Thomas. *American Museum of Natural History Anthropological Papers,* in press.

WILD, C. J., and R. K. NICHOL
 1983 Estimation of the original number of individuals from paired bone counts using estimators of the Krantz type. *Journal of Field Archaeology* **10**:337–344.

WILSON, R. W.
 1960 Early Miocene rodents and insectivores from northeastern Colorado. *University of Kansas Paleontological Contributions, Vertebrata* **7**:1–92.

WING, E. S.
 1963 Vertebrates from the Jungerman and Goodman sites near the east coast of Florida. *Florida State Museum Contributions, Social Sciences* **10**:51–60.
 1975 Hunting and herding in the Peruvian Andes. In *Archaeozoological*

studies, edited by A. T. Clason. Amsterdam:North-Holland. Pp. 302–308.

WING, E. S., and A. B. BROWN
   1979 *Paleonutrition: Method and theory in prehistoric foodways.* New York: Academic Press.

WITTRY, W.
   1959 The Raddatz Rockshelter, SK5, Wisconsin. *Wisconsin Archaeologist* **40**:33–69.

WOLFF, R. G.
   1975 Sampling and sample size in ecological analyses of fossil mammals. *Paleobiology* **1**:195–204.

WOODWARD, J.
   1728 *Fossils of all kinds, digested into a method, suitable to their mutual relation and affinity.* London.

Index

A

Abundance
  absolute, 17, 30, 32-34, 37-40, 86-88
  relative, 17, 25, 26, 30, 32-34, 38-40, 43-44, 50, 63, 67-68, 86-88, 92, 96, 116-130
Aggregate, faunal
  defined, 17
Aggregation, faunal
  effects of, 29-49, 50, 51, 63-68, 83-85, 90-92, 94-95, 98-99, 100-101, 106, 173
Alder, G. M., 118-119
Apple Creek, 54, 55, 63, 72, 91, 97
Assemblage, faunal, 25
  defined, 17

B

Backhoe Village, 146, 148-149, 157, 162, 164
Bate, D. M. A., 18-20, 25
Bear River 1, 146, 148-149, 153-157, 162, 163, 164, 165, 167
Bear River 2, 146, 153, 157, 162, 164
Bear River 3, 146, 153-157, 162, 163, 164, 165, 167
Bedwell, S. F., 34
Benson, S. B., 5
Binford, L. R., 20, 88-89, 96
Biomass, 17, 22, 67, 172
Black Rock III, 146, 161
Boardman fauna, 77-80
Bone preservation, 21-22, 25, 26, 28, 49, 87
Brain, C. K., 2, 21, 25, 45, 49, 67
Buffalo, 54, 56, 63, 69-70, 72, 73-75, 91, 98, 102, 103, 104
Butchering, 20-21, 26, 27, 68, 73, 105, 149-150, 154, 164
Butler, B. R., 121

C

Caldwell Village, 146, 162, 164
Casteel, R. W., 29, 51, 52, 53, 62, 70, 72, 87-88, 168-173
Chaplin, R. E., 23, 68, 173
Clark's Cave, 58-61, 67
Cleland, C. E., 120-121, 127, 131
Closed arrays, 19, 159
Collection techniques, 22, 25, 168-172
Connley Cave No. 4, 34-40, 41, 64-66, 81, 83-84, 85, 86, 91, 99-101, 103
Connley Caves, 112-113, 114, 121, 123, 124
Cultural bone, 80-81
Cumulative distribution functions, 154-158

D

Daly, P., 20-21
Davis, J. O., 4, 12
Dirty Shame Rockshelter, 54, 57, 58, 62, 63, 72, 99, 100
Diversity, taxonomic, 131, 152
  indices, 152, 158-167
  sample size effects, 158-167
  species-abundance distributions, 152-158
Ducos, P., 50-51, 52, 53, 67, 68
Dunnell, R. C., 131

E

Elements
  defined, 16
  estimated number of, 77

minimum numbers of (MNE), 90
most abundant, *see* Most abundant element
Evans Mound, 146, 162, 163, 164, 165

F

Felter Hill, 146, 149
Fieller, N. R. J., 87–88
Flannery, K. V., 28, 35
Fort Ligonier, 54, 59, 63, 72, 97, 101, 104–105, 106, 111
42SL19, 146, 161
Fremont culture
  avian assemblage diversity, 160–167
  avian assemblage richness, 143, 145–147
  general description, 141–142
  mammalian assemblage diversity, 153–158, 160–167
  mammalian assemblage richness, 145–151
  sites, excavation of, 171
  subsistence, 143, 153
Furlong, E. L., 5

G

Gatecliff Shelter, 3–5, 6–11, 91, 113–114, 121–124, 125, 130, 132, 138, 139, 140, 144, 171
Gejvall, N.-G., 71–72, 73
Gifford, D. P., 52, 75, 105, 106
Gilbert, A. S., 26
Greene, J. L., 117–118
Guilday, J. E., 1, 58–61, 67, 75, 104–105, 111, 114–115
Guthrie, R. D., 21

H

Hanging Rock Shelter, 81, 82
Harper, K. T., 118–119
Hemphill fauna, 77–80
Hesse, B., 26, 55–57, 58, 61, 66, 67
Hidden Cave, 3, 5, 12–15, 34, 40–49, 66, 91, 107–108, 109, 115, 124–127, 130, 132, 135–137, 138–139, 141, 142, 143, 144, 171
Hogup Cave, 118–120, 121
Hohokam, 117–118
Honerkamp, N., 173
Howard, H., 5

I

Identifiability, differential, 21, 25, 77
Injun Creek, 146, 162, 163, 164, 165, 167

Innocents Ridge, 146, 161
Interdependence of identified specimens, 23–24, 25–26, 28, 41, 49–50, 66–68, 99, 106

J

Jungerman, 159–160, 161

K

Klein, R. G., 90
Knoll, 146, 162, 163, 165, 167
Krantz, G. S., 86–88
Kromdraai, 179

L

Lahontan Basin, 5, 40, 126
Lerna, 71–72
Levee, 146, 162, 163, 165, 167
Little Smoky Creek Shelter, 81, 82
Lyman, R. L., 21, 25, 173

M

MacArthur, R. H., 160
McKay fauna, 77–80
Madsen, D. B., 153, 161, 166
Makapansgat, 45
Matched-pairs methods, 86–88
Matthews, T. W., 117–118
Maximum number of individuals, 96
Mayer, D., 91
Meadowcroft Rockshelter, 139–140, 142, 144, 145
Measurement scales
  interval, 93
  ordinal, 93, 96–111, 130
  ratio, 93, 94, 95–96, 110
Meat weights, 17, 22–23, 24, 27, 104–105, 172–174
Median Village, 146, 149–151, 157, 162, 164
Mehringer, P. J., Jr., 40, 124, 125
Minimal animal units (MAU), 88–90
Minimum number of individuals (MNI), 20, 23, 24, 25, 27–49, 76–77, 85–88, 88–89, 90–92, 117, 119, 130, 151, 152, 153, 157, 172
  diversity measures using, 158–167
  as ordinal scale measure, 96–111, 130
  as ratio scale measure, 95–96, 110
  relationship to MNI/NISP, 68–84
  relationship to NISP, 49–68
  relationship to NISP/MNI, 68–84

INDEX

Morrison, R. B., 5
Most abundant element (MAE), 29–34, 36, 38–39, 43, 44, 49, 58, 64, 66, 88–89

N

Natural bone, 80–81
Nephi, 146, 162, 163, 164, 165
Nichol, R. K., 87–88
Noddle, B. A., 173
Number of identified specimens (NISP), 17–26, 28, 30, 31, 32, 36–37, 40, 41, 44, 65–66, 85–88, 89, 90–92, 117, 121–130, 152–153, 157, 172; *see also* Specimen
  diversity measures using, 158–167
  as ordinal scale measure, 96–111, 130
  as ratio scale measure, 94–96, 110
  relationship of MNI/NISP to MNI, 68–84
  relationship of NISP/MNI to MNI, 68–84
  relationship to MNI, 49–68
  relationship to taxonomic richness, 132–151

O

Old Woman, 146, 162, 164

P

Paaver, K., 29
Parmalee, P. W., 1, 55–57, 120, 143–144, 161
Payne, S., 172
Percentage survival of skeletal parts
  aggregation effects on, 44–49
Perkins, D., 20–21, 26
Pharo Village, 146, 149–151, 157, 162, 163, 164, 165
Poplar Knob, 146, 161
Prolonged Drift, 52, 53, 54, 63, 70–71, 72, 73, 75, 76, 81, 102, 104, 105, 106, 110
Proximal community, 75–76, 77, 79–80

R

Raddatz Rockshelter, 120–121, 127–129, 130
Rancho La Brea, 27–28
Rank order invariance, 106–110, 115
Rarefaction, 151–152
Reitz, E., 173
Relative skeletal completeness, 75, 77, 80, 81
Residuals, 51, 52, 80, 81, 83, 147–151
  adjusted, 153–154, 155
Richness, taxonomic, 131, 132–151

S

Sample size, 37, 43, 44, 50, 79, 112, 151–152, 157
Sample size effects on
  artifact class relative abundance, 179
  artifact class richness, 179
  diversity measures, 158–167
  relative abundance, 116–130
  taxonomic richness, 132–151
Sanders, H. L., 151–152
Schiffer, M. B., 179
Schlepp effect, 20–21, 105
Screen size
  effects of, on bone recovery, 4, 22, 168–172
Seasonal dating, *see* Seasonality
Seasonality, 174–177
Shannon index, 159–160
Shotwell, J. A., 2, 16, 21, 25, 68, 72, 73, 75–80, 81
Significance tests
  aggregation effects on, 38, 39–40, 43–44
  interdependence effects on, 41
  sample size effects on, 22–23, 44
Simpson's index, 160
Smirnov test, 152, 154–158
Smith, B. D., 132
Smoky Creek Cave, 81, 82
Snake Rock, 146, 162, 164
Snaketown, 117–118
Specimen, *see also* Number of identified specimens
  corrected number, per individual (CSI), 76, 77–80, 81
  corrected number, per taxon, 76–77
  defined, 16
  identified, 17
  unidentifiable, 16
Steinfeld, P., 26
Sterkfontein, 179
Styles, B. W., 152
Swartkrans, 49, 91, 179

T

Tabūn Cave, 18–20
Taphonomy, 1, 2, 26, 34, 88, 105, 106, 108, 115, 151, 158, 160, 171, 180
Thomas, D. H., 4, 12, 68, 80–81, 91, 168–171
Thompson, R. H., 4
Tipper, J. C., 152, 156
Turner, A., 87–88

V

Vaufrey, R., 20
Voorhies, M. R., 2

W

Wad Cave, 18-20
Warren, 146, 162, 163, 165, 166
Wasden, 121, 122, 123

Westend Blowout fauna, 77-80
White, T. E., 1, 20, 27-28, 29, 35, 172
Whiterocks, 146, 161
Wigand, P. E., 40, 124, 125
Wild, C. J., 87-88
Windy Ridge, 146, 162, 164
Wing, E. S., 159-160, 161, 163
Woodward, J., 174